Geographies of Food and Power

This book provides a comprehensive overview of the production and consumption of food, suitable for use in undergraduate classrooms, either at the intermediate or advanced level.

It takes an intersectional approach to difference and power and approaches standard subjects in the geography of food with a fresh perspective focusing on inequality, uneven production and legacies of colonialism. The book also focuses on places and regions often overlooked in conventional narratives, such as the Americas in the domestication of plants. The topics covered in the textbook include:

- descriptions and analyses of food systems
- histories of agricultural development with a focus on the roles of different regions
- major commodities such as meat, grains and produce with a focus on the place of production
- contemporary challenges in the food system, including labor, disasters/conflict and climate change
- recent and emerging trends in food and agriculture such as lab-grown meat and vertical urban farms

Geographies of Food and Power takes a synthetic approach by discussing food as something produced within an interconnected system, in which labor, food quality and the environment are considered together. It will be a valuable resource for students of human geography, environmental geography, economic geography, food studies and development.

Amy Trauger is Professor of Geography in the Department of Geography at the University of Georgia.

"This is the geography textbook for the class I always wished I'd taken, progressing from introductory concepts to sophisticated analysis over the arc of a semester. With terrific suggestions for supplementary reading, watching, and discussion, Amy Trauger's *Geographies of Food and Power* is set to become a classic foundation for generations of geographers."

Raj Patel,
Research Professor, Lyndon B. Johnson School of Public Affairs,
The University of Texas at Austin, USA

"*Geographies of Food and Power* makes a critical contribution to ongoing debates about the historical and contemporary structures of our food system along with the social and environmental implications. Using an intersectional lens, Trauger introduces several concepts and theoretical perspectives to reveal challenges and point to promising pathways forward. This is an essential text for students of geography and food studies along with anyone interested in just and sustainable food futures."

Charles Z. Levkoe,
Canada Research Chair in Equitable and Sustainable Food Systems,
Lakehead University, Canada

Geographies of Food and Power

Amy Trauger

LONDON AND NEW YORK

Cover image: © Amy Trauger

First published 2023
by Routledge
4 Park Square, Milton Park, Abingdon, Oxon OX14 4RN

and by Routledge
605 Third Avenue, New York, NY 10158

Routledge is an imprint of the Taylor & Francis Group, an informa business

British Library Cataloguing-in-Publication Data
A catalogue record for this book is available from the British Library

Library of Congress Cataloging-in-Publication Data
A catalog record for this book has been requested

ISBN: 978-0-367-74152-5 (hbk)
ISBN: 978-0-367-74766-4 (pbk)
ISBN: 978-1-003-15943-8 (ebk)

DOI: 10.4324/9781003159438

Typeset in Berling and Futura
by Apex CoVantage, LLC

Access the Companion Website: www.routledgetextbooks.com/textbooks/9780367747664

Contents

About the Author

Amy Trauger is Professor of Geography in the Department of Geography at the University of Georgia. She is Affiliate Faculty with the University of Georgia's Institute for Women's Studies, the Latin American and Caribbean Studies Institute and the Center for Integrative Conservation Research. Her research interests include food sovereignty, sustainable and alternative agriculture, human-environment interactions, gender and agriculture, and Indigenous political struggle. She taught the Geography of Food and the Athens Urban Food Collective courses at the University of Georgia nearly every year since she arrived in 2008. Before that, she taught the Geography of Sustainability course at Penn State University. She has written, edited and collaborated on four books on the topics of food, agriculture and the environment. Her book *We Want Land to Live* (2017) is used in graduate and advanced undergraduate classes in the United States. Her work is widely used by professors teaching courses on the geography of food and agriculture at Iowa State University, West Virginia University, Tufts University, Cornell and Washington State, among others.

Figures

PART 1

Food and Power

CHAPTER 1

The Geographies of Food

INTRODUCTION

At a conference in Bristol, England, in June 2005, a colleague and I began discussing a research project focused on the rapid increases in the globalization of organic food. During a break in the conference, we made our way to several local food shops and spent a few hours examining the geographic origins of the produce in the UK. I noticed that the vast majority of organic and fair trade bananas for sale were sourced from the Dominican Republic. As a then frequent banana eater, foodie and geographer, I knew that organic bananas for sale in US markets were sourced from Colombia, Honduras and Guatemala. I did not know that the Dominican Republic was one of the largest suppliers of organic and fair trade bananas to the UK and Europe. Figure 1.1 shows a Haitian banana worker packing fair trade bananas to be shipped to Europe. Given that the Dominican Republic was much closer to the US than those other sources, this discrepancy presented a classic geographic question. If distance increases costs, why would a closer source of organic bananas be passed over for a more distant one?

The answer is rooted in imperialism and the way political-economic relationships shrink the costs of distance. The UK sourced its bananas from the Caribbean, according to preferential trade agreements in the wake of independence movements in their former colonies. European ex-patriots who settled in the Caribbean dominated trade, including that of fair trade and organic bananas. Meanwhile, the proxy wars the US fought in the early to mid-20th century in Latin America and the coups they supported paved the way for American corporations, such as the precursors to Dole and Chiquita, to dominate trade to the US from those countries. The owners of long-established banana plantations in Colombia, Guatemala and Honduras sectioned part of their conventional production into certified organic production and exploited their long-standing trade relationships with the US to add organic products to their portfolio. Thus, contemporary production of products marketed for their attributes of social and environmental justice follows patterns of power and dominance that were established through colonialism and imperialism, including the use of migrant labor who are forced out of their homes by neo/imperialism.

This chapter is a comprehensive introduction to geographic themes and concepts used to discuss food and power. In short, power works through the food system along paths created by imperialism, with negative outcomes for human health and the environment that are most impactful on the poor, migrants, people of color and women. Sustainability

DOI: 10.4324/9781003159438-2

FIGURE 1.1 A Haitian worker in the Dominican Republic

prioritizes social justice, economic viability and environmental health in the contemporary food system, but due to its embeddedness in the historical and contemporary geographies of power and capitalism, it often falls far short of delivering on that promise. In what follows, I discuss histories of colonialism and their relationship to agriculture and the global food system.

THE GEOGRAPHIES OF FOOD

The stories of food follow in the footsteps of empire. Access to food was a central objective of all imperial expansions; tea, sugar, salt and spices have long been traded through imperial networks. Through their trade, these products became known as agricultural **commodities**, which refers to any item that is destined to be bought and sold. In the age of European imperialism, they became staples of European diets and tools of an empire after the colonization of vast areas of the planet. The land, labor and resources of the so-called New World produced the nutritional advantages, the luxuries, and the wealth from manufacturing that Europeans enjoyed at the expense of the colonized who were dispossessed of their land and resources, enslaved and murdered by the colonizers if they resisted (Pascoe, 2014). The modern food system inherited this legacy and continues to reflect its history in spite of attempts at reform. **Neoimperialism** in the form of proxy wars, land grabs and corporate-controlled agriculture continues to shape the food system and what appears on our plates, from where and who pays for it and in what ways.

Agriculture is a complex system of human and natural components, including land, water and seeds, that are brought together through labor and technology to produce food. Figure 1.2 demonstrates a general pattern of agricultural production and modernization. The application of technology, such as tools or fertilizers, to agriculture increases the efficiency of food production, allowing people to do other things besides grow or gather food. This **specialization of labor**, very often waged, results in the **migration** of people to settlements to find work, which in turn grow into cities. **Urbanization** is accompanied by rapid population increase, which results in an increased demand for food, requiring new combinations of land, seed, water, technology and labor. When the resources (soil fertility, water)

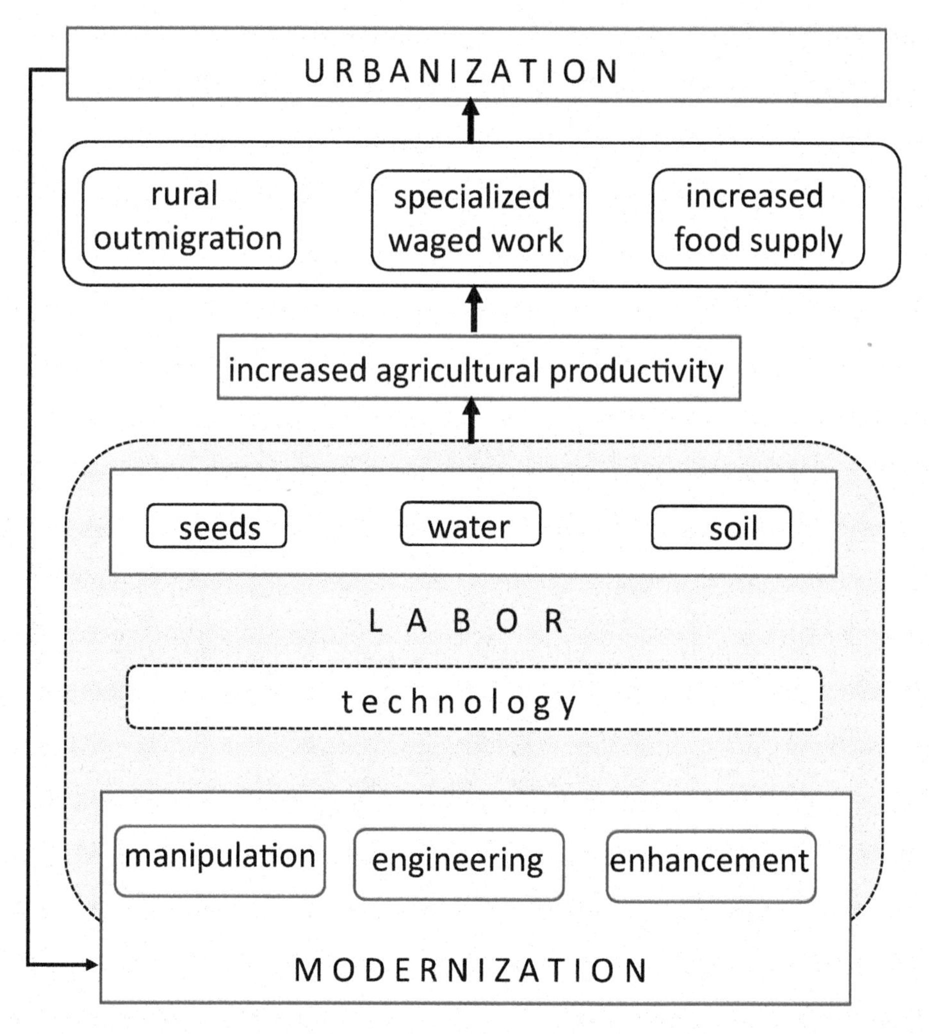

FIGURE 1.2 Modernization processes in agriculture (adapted from Marston & Knox, 2007)

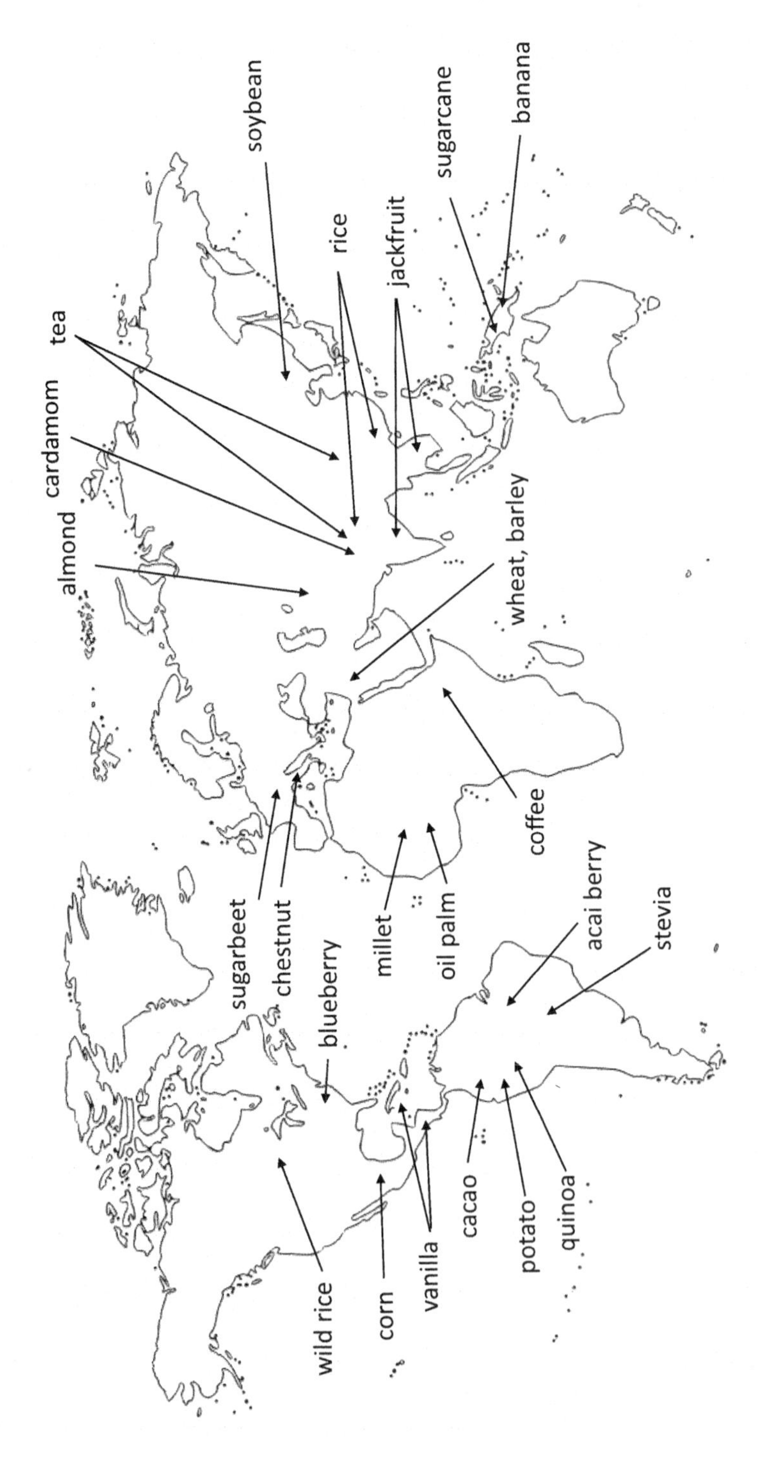

FIGURE 1.3 Centers of origin for selected foods discussed in this book

in one site are exhausted, food producers, whether farmers, colonizers or corporations in a capitalist system, will move elsewhere, in something called the **spatial fix** (Harvey, 1990), creating extended networks that shape the dynamics of our contemporary food system.

As European empires grew spatially and geopolitically in the 1400s, exhausting resources within their territorial boundaries, explorers sought out new land to colonize in other places. The resulting Age of European Colonial Terrorism (Jalata, 2016) included the dispossesion of Indigenous people of their land, the exporting of their food crops and seed stock back to the **metropole** and the enslavement of Africans and Indigenous people to produce food and valuable commodities for Europeans and settlers. Figure 1.3 shows the origin of food crops discussed in this book, showing how foods that are ubiquitous in supermarkets around the world were once (and perhaps still are) staple foods for peasants and Indigenous people everywhere. Imperial organization of the pathways for food to travel not only dispossessed and enslaved people to work on plantations of sugar, tea and coffee (Mintz, 1986), it also created a system of **dependency** that often persists after independence.

After independence, which for many colonies was in the 1940s–1960s, the political-economic relationships around food production and export remained in place. **Dependency theory** asserts that current or former colonies are trapped in a situation where colonies have a product, such as bananas, and are required to sell a certain amount (quotas) to certain countries at certain prices, often fulfilling these mandates at the expense of their own autonomy and prosperity (Ferraro, 2008). This results in the development of **single-resource economies**, or those countries whose economic development is tied and often limited to the production and export of a single commodity, such as bananas, tea or sugar. Smith (2010) refers to the ongoing production of inequitable economic and social relationships between places, including colonizing countries and former colonies as **uneven development**.

FOCUS SECTION: POTATOES

Potatoes were domesticated in the Andes mountains in what is now southern Peru and northern Bolivia approximately 7,000–10,000 years ago. Potatoes were adapted by Indigenous people from the starchy roots of a wild potato species, and through trade between Indigenous people spread to south-central Chile. While more than 4,000 varieites of potatoes now exist, nearly all of them are descended from cultivars further developed in Chile. A key staple of the Incan diet, potatoes remain an important source of carbohydrates for Indigenous people throughout the Andes. Spanish imperialists exported potatoes to North America, Europe and Asia, where they were readily adapted to local diets due to their high yield and easy cultivation. Potatoes are grown from either a sprout on the root (thereby making it a clone of its parent), which is very easy, or from seed, which takes much longer. The potato contributed to population growth and urbanization in Europe between 1700 and 1900, but its vulnerability to a fungus when grown from root sprouts led to the Great Irish Famine beginning in 1845. The economy and population in Ireland collapsed and triggered significant outmigration to the Americas. In spite of the abundance of cultivars, the vast majority of people today eat only a few varieties, such as the Russet Burbank, which is largely determined by their durability while being transported over long distances and cultivated mass-market tastes, such as those for french fries.

Production-Consumption Networks

Food is distributed from producers to consumers in a complex web of interrelated places and processes known as a **food system**. This web is composed of nodes of *production, distribution* and *consumption* through which food is processed, delivered and marketed to eaters. It also includes food waste and the natural resources required to produce the food, such as land, water and seeds. Other words for food systems include commodity chains or **supply chains**, which trace a product from raw material to consumable good. A supply chain specifically refers to a commodity chain that is shaped largely by the supplier, which in the case of processed food is a food manufacturer. I use the concept of the supply chain to capture the power dynamics of the food system, which are tightly controlled by those who handle the transformation of raw materials and distribution of finished food products. Figure 1.4 shows a simplified food system structured in the form of a supply chain. Generally, many producers deliver raw materials or commodities into a relatively small distribution network that will have its own internal supply chain of purchasers, processors and sales, and the finished products are then delivered to many consumers via retail or wholesale outlets such as grocery stores and restaurants.

All food systems are characterized by some form of distance between eaters and farmers. The concept of **food miles** (Pretty et al., 2005), or the physical distance that one food

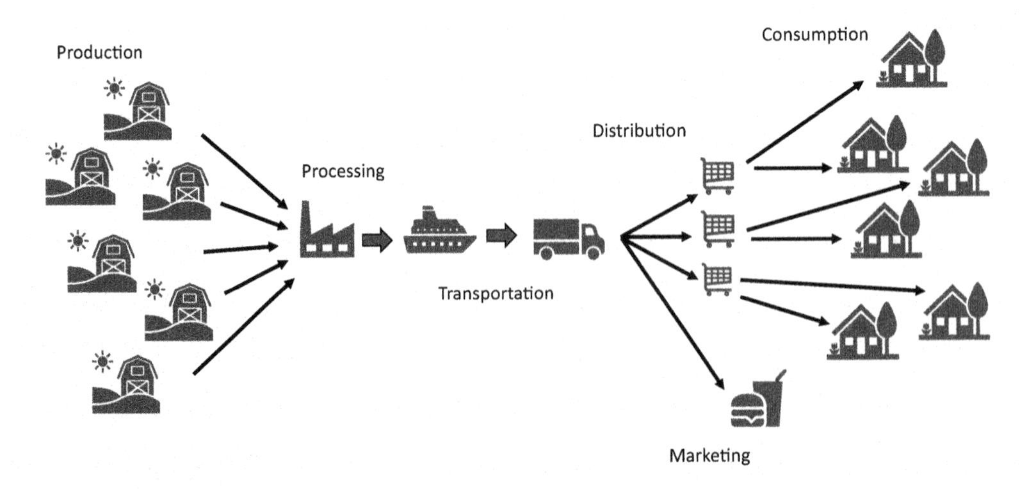

FIGURE 1.4 Simplified food system in the form of a supply chain

item travels, attempts to capture this distance. It has been said that the average food bite travels 1,500 miles from farm to fork in the US. In addition to physical distance, food systems are characterized by what sociologists call **social distance** or the mental distinctions people create between people of different countries, races or social classes (Wark & Galliher, 2007). Both the physical and the social distance between the farmers and laborers who produce food and the eaters who consume it generate *inequalities* and barriers to knowledge about the social, economic and environmental challenges that confront the contemporary food system. These include threats to *human and environmental health*, often in the form of pollution. These distances also contribute to inequalities between different social groups, specifically in terms of access to food for poor people and access to resources for farmers in developing countries. Both of these situations present challenges to the *sustainability* of the contemporary food system. These three themes are carried forward throughout the book, and I examine each in more detail in the following sections.

MAJOR THEMES

Sustainability

Sustainability is often referred to as the triple bottom line of social justice, economic viability and environmental health (Whitehead, 2007). Sustainable development was a term coined in 1987 in the Brundtland Report to the United Nations on economic trajectories for the 21st century. It encouraged nation-states to pursue economic development strategies that met the needs of the present generation but did not impede the ability of future generations to access resources. Sustainability as a concept has received a significant amount of critique from multiple directions, but it nevertheless persists as a goal for governments and non-governmental organizations to address the problems of inequality and environmental damage caused by agriculture. Figure 1.5 illustrates how sustainability lies at the intersections of efforts to protect the environment, secure profitability and enable high standards of living for all, which manifest in worker's rights, reductions in pollution and more effective uses of natural resources, such as sunlight. The approaches taken to solving the problems of the food system vary by scale, but those that have found the most success are at small community scales with strategies that are attuned to the local environment and to the needs of specific populations in geographically distinct places.

Health and Environment

The threats that unsustainable agriculture poses to **human and environmental health** are well documented, particularly since the 1962 publication of Rachel Carson's *Silent Spring*. Carson called attention to the threats to human and animal health from toxic pesticides and demonstrated how human and natural health are linked. While some national-scale governments have done much to limit the production and distribution of toxins in the food system, problems persist, including and especially from insecticides and fungicides that are used globally to produce food destined to be consumed where such substances may be banned. Workers are exposed to them in the field, and consumers are exposed

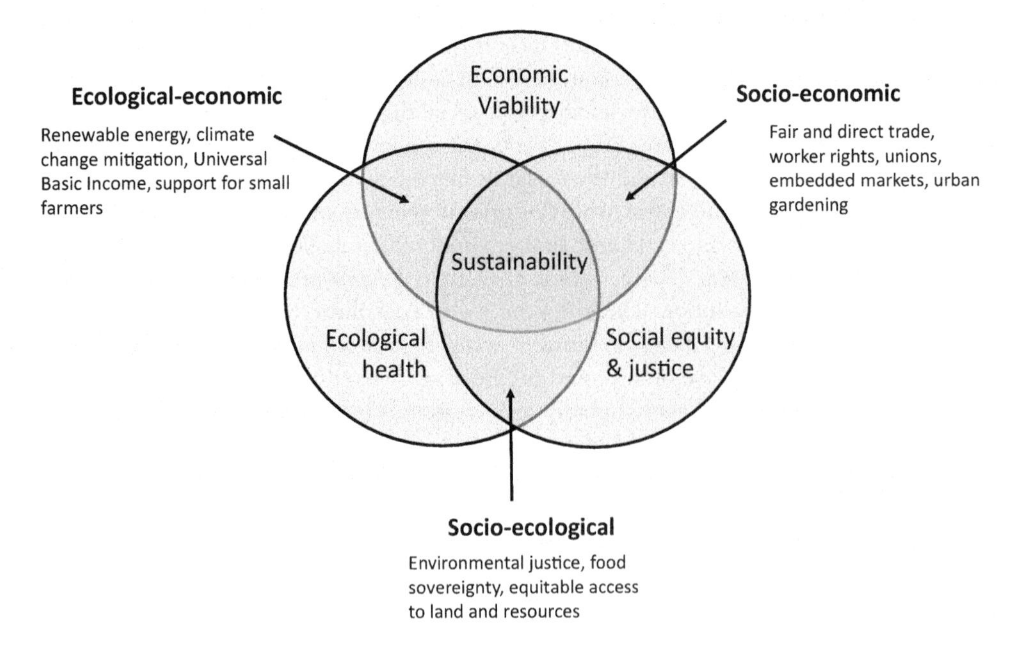

FIGURE 1.5 The three spheres of sustainability (adapted from Abubakar & Attanda, 2013)

when they eat the food. Plasticizers and other food additives also lead to adverse health effects, and research often cannot keep up with the proliferation of chemicals and their myriad effects on humans and the environment (Guthman, 2011). Industrial food systems contribute an estimated half of all greenhouse gas emissions, making them a major driver of climate change as well as an industry deeply affected by it through shifting zones of production, catastrophic storms, flooding and drought, with disastrous impacts in place-specific ways. Lack of access to food in one place and surpluses in another constitute a major contradiction of the contemporary food system with widely varying health outcomes for differently situated populations. Figure 1.6 graphically demonstrates the many ways agriculture impacts human and environmental health.

Inequality and Intersectionality

Deep disparities in the production and distribution of food contribute to the paradox of overnutrition in one place and undernutrition in another. The conventional wisdom that we do not have enough food to feed a growing population breaks down when scholars examine surplus and waste in wealthy countries with the opposite problems of hunger and malnutrition in poor countries. This is a contradiction largely driven by inequalities in distribution, which is a function of global capitalism and the legacies of colonialism. Some of the most impoverished and malnourished people are, paradoxically, farm laborers who produce food but cannot afford to access it. In wealthy countries, hunger is often associated with lack of access to nutritious food for various reasons and is often a result of **intersectional inequalities**, the interlocking and overlapping oppressions of race, class

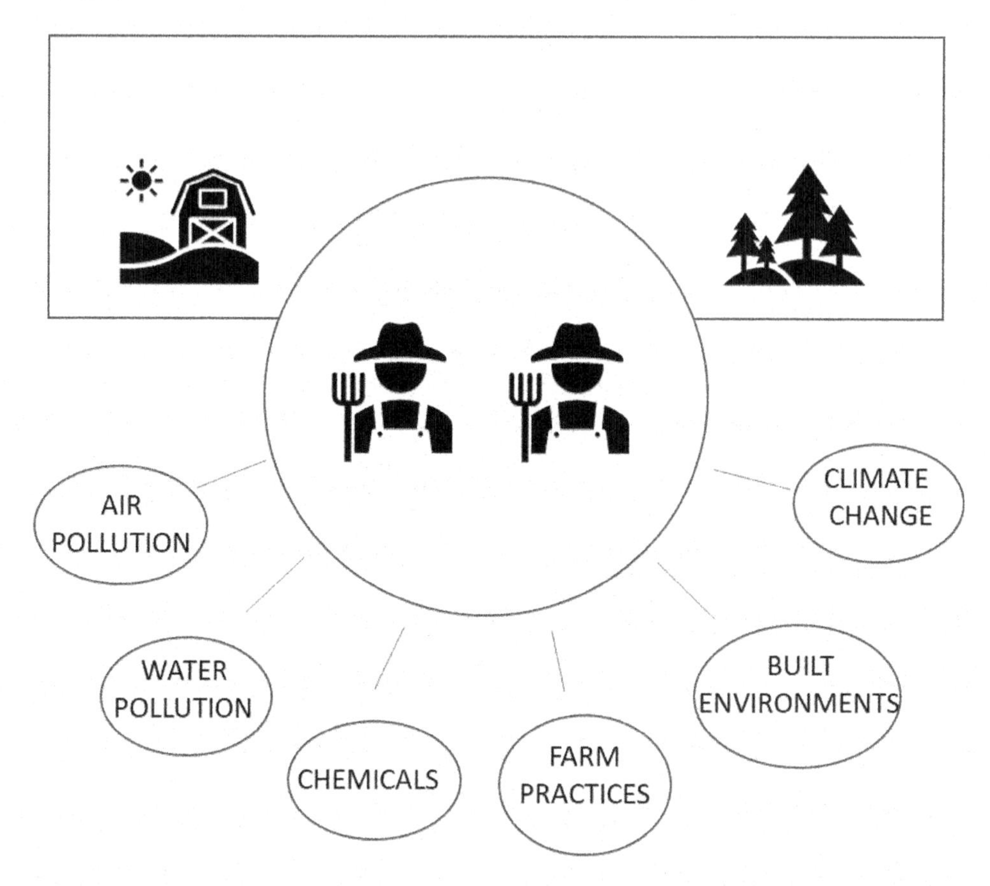

FIGURE 1.6 The intersections of human and environmental health (adapted from Brusseau et al., 2019)

and gender (Crenshaw, 1991). There are deep inequalities with regard to access to land in many countries, with wealthy landowners monopolizing production of global commodities, often with the help of governments that subsidize production for economic and geopolitical reasons that have little to do with feeding people. Four central categories of identity used to marginalize people in the food system are gender, race, class and citizenship status. Figure 1.7 shows some, but not all, categories of identity that are often imposed on others and used to justify mistreatment. Where they intersect is a locus of multiple identities that may prevent food system actors from participating fully, thus disproportionately experiencing hunger, exploitation and exclusion.

CHAPTER OVERVIEWS

Chapter 2 provides a basic introduction to geographic thinking and outlines five key geographic concepts used throughout the book: scale, distance, region, place and networks.

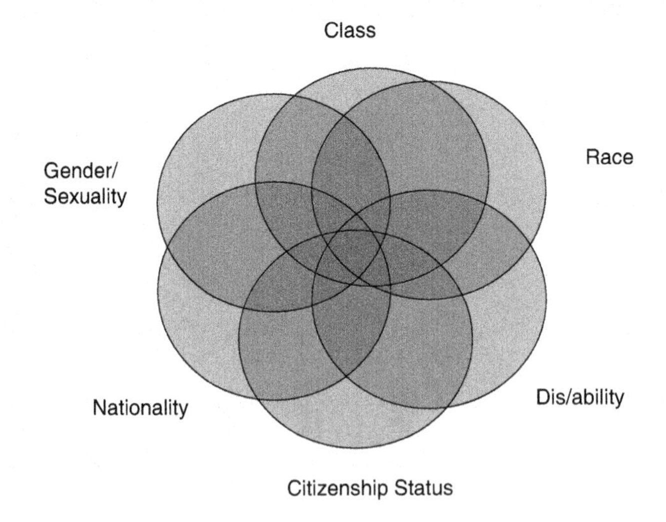

FIGURE 1.7 Intersecting and overlapping axes of identity

I discuss them in this chapter and throughout the text as they relate to the way geopolitical and political-economic power impact the food system. This chapter also elaborates on the three central themes identified in Chapter 1 and that are followed throughout the book.

Chapter 3 begins a section on food system politics and provides an overview of the changes in agriculture that have shaped human societies. These include Indigenous science and methods, Mendelian genetics, chemical applications and bioengineering. Chapter 4 provides a comprehensive overview of food systems and their associated vocabularies that are used throughout the book. This chapter underscores the integrated approach to studying the creation and governance of supply chains from field to fork and beyond through the lens of geography. Chapter 5 elaborates on the relationship between inequities and injustice in the food system that have received scholarly attention in the late 20th century and the early 21st century. This chapter focuses on the contradictions of the modern food system that produces too much food in one place or region and not enough or not the right kind in another, thus generating surpluses and waste on one hand and hunger and malnutrition on the other. Chapter 6 explores the social and political responses to hunger by examining the right to food, food justice, food sovereignty, local agriculture and sustainability and their place-specific challenges and successes.

Chapter 7 begins a section of the book focusing on different types of commodities and their supply chains. This chapter includes summaries of broad categories of fresh fruits and vegetables and their places of production, the (neo)colonial relationships, labor relations and technological innovations that continue to shape supply chains. Chapter 8 discusses grains and seeds as staple carbohydrates and pulses and nuts as vegetarian sources of protein. The chapter discusses mechanization, economies of scale and the role of cheap food as a foundation for a state's economy. Chapter 9 discusses the production of animal proteins throughout the world with sections on meat, fish, dairy products and eggs. This chapter covers the practices, technological innovations and governance associated with animal protein production across the world with an emphasis on alternatives, ethics and sustainability. Chapter 10 focuses on tropical commodities that are produced as luxury goods in former colonies for their former colonizers and/or consumers in wealthy nations as part of the process of modernization. This chapter focuses on coffee, cacao and sugar as historically significant commodities with contemporary challenges with regard to labor, climate change and food insecurity and ongoing uneven development. Chapter 11 provides an overview of synthetic and natural ingredients and additives used to enhance flavor, color or texture, some of which pose threats to human health. While there is a dizzying array of additives, this chapter discusses three in detail: sweeteners, preservatives and fats.

Chapter 12 introduces a section on the interconnected contemporary challenges to agriculture and food systems: labor, conflict and disasters and climate change. This chapter provides a basic geography of agricultural labor, governance of labor conditions, including social movements for fair working conditions, the use of migrant labor and the challenges to provide equity for all workers in the food system. Chapter 13 focuses on historic and contemporary challenges to food supply chains resulting from natural disasters and geopolitical conflict, with an emphasis on famine during wartime and food scarcity issues and factors influencing supply, such as panic buying and labor shortages. Chapter 14 focuses on the slowly unfolding disaster of climate change and the unprecedented challenges it presents to food systems. This chapter elaborates on the food system contributions to climate change via greenhouse gas emissions and major threats to agriculture posed by climate change, including flooding and disruptions to food supply chains. This chapter also discusses alternative fuels, new production models and mitigation strategies. The last chapter of this section includes a discussion of future trends and summarizes the connections between capitalism, labor, environment and inequality and how governance of the food system can help or hinder progress toward equity, sustainability and health.

SUMMARY

This chapter introduced some key concepts and terms that are used throughout the book. Primary among them is the way in which both the problems and products of agriculture proceed from innovations, technology and the expansion of agriculture's spatial reach. In the Age of European Colonial Terrorism, food distribution networks expanded globally, a process that continues in the form of extended supply chains, which bring with them foods from all over the world to new markets, but also pollution, disease, inequality and unsustainability. The relationship of food and power to geography is the subject of the following chapter.

SUPPLEMENTAL MATERIALS

Key Terms and Concepts: agriculture, commodities, dependency theory, food miles, food system, human and environmental health, intersectional inequalities, metropole, migration, neoimperialism, single-resource economies, social distance, spatial fix, specialization of labor, supply chains, sustainability, uneven development, urbanization

Explore More: The Taste Atlas is an interactive website where users can explore thousands of different kinds of foods from around the world. www.tasteatlas.com/

Recommended Viewing: *Our Daily Bread* (2005); *Seed: the Untold Story* (2016)

Discussion Prompt: What is the historical geography of a commodity you frequently consume? (a) Identify it (wheat, coffee, bananas, etc.) and tell us why/when/how you eat it, (b) locate its origin of domestication by identifying a modern country and/or an Indigenous people/nation and (c) describe its global journey to the modern diet. Aim for 100–250 words. Be sure to correctly use at least *one concept* from this chapter in your discussion post and to write it in **bold** font.

Recipe: Milcaos is a traditional Chilean recipe for potatoes from Chiloé, the agricultural hearth for potato varieties today. Traditionally, only Chilotéan potatoes should be used, but any red potato can be used. www.chileanfoodandgarden.com/chilean-potato-bread-chapaleles/

REFERENCES

Abubakar, M. S., & Attanda, M. L. (2013). The concept of sustainable agriculture: Challenges and prospects. In *IOP conference series: Materials science and engineering* (Vol. 53, No. 1, p. 012001). IOP Publishing.

Brusseau, M. L., Ramirez-Andreotta, M., Pepper, I. L., & Maximillian, J. (2019). Environmental impacts on human health and well-being. In *Environmental and pollution science* (pp. 477–499). Academic Press.

Carson, R. (1962). *Silent spring*. Houghton Mifflin Harcourt.

Crenshaw, K. (1991). Mapping the margins: Intersectionality, identity politics, and violence against women of color. *Stanford Law Review*, 1241–1299.

Ferraro, V. (2008). Dependency theory: An introduction. *The Development Economics Reader*, *12*(2), 58–64.

Guthman, J. (2011). *Weighing in: Obesity, food justice, and the limits of capitalism* (Vol. 32). University of California Press.

Harvey, D. (1990). *The condition of postmodernity: An enquiry into the conditions of cultural change*. Blackwell.

Jalata, A. (2016). Colonial terrorism and the incorporation of Africa into the capitalist world system. In *Phases of terrorism in the age of globalization* (pp. 87–113). Palgrave Macmillan.

Marston, S., & Knox, P. (2007). Places and regions in global context. In *Human geography*. Pearson Prentice Hall.

Mintz, S. W. (1986). *Sweetness and power: The place of sugar in modern history*. Penguin.

Pascoe, B. (2014). *Dark Emu black seeds: Agriculture or accident?* Magabala Books.

Pretty, J. N., Ball, A. S., Lang, T., & Morison, J. I. (2005). Farm costs and food miles: An assessment of the full cost of the UK weekly food basket. *Food Policy*, *30*(1), 1–19.

Smith, N. (2010). *Uneven development: Nature, capital, and the production of space*. University of Georgia Press.
Wark, C., & Galliher, J. F. (2007). Emory Bogardus and the origins of the social distance scale. *The American Sociologist, 38*(4), 383–395.
Whitehead, M. (2007). *Spaces of sustainability: Geographical perspectives on the sustainable society*. Routledge.

CHAPTER 2

Geography and Power

INTRODUCTION

In the example that opened the introductory chapter, organic bananas traveled farther from their closest markets, which was explained by preexisting economic relationships rooted in colonialism. What that did not explain, however, was that one of the largest markets for organic bananas produced in the Dominican Republic is Japan. The US buys most of its bananas, including the organic and fair trade kind, from Colombia, Honduras and Guatemala. Figure 2.1 shows shipping containers in San Diego delivering bananas from Central America. If **distance**, or the physical space between two places, increases the costs of transportation, which is a fairly straightforward geographic and economic assumption, why would it be cost-effective or even profitable to send a relatively low-value product halfway around the world? And why would the exporters in the Dominican Republic ship bananas to Japan when they could sell them to a market in the US, a mere 1,000 miles away?

The answer to why it is cheaper to ship bananas to Japan than to almost anywhere else, including the US, lies in an empty shipping container. Global electronics manufacturers, some of whom are located in Japan, ship semifinished goods to the Dominican Republic for finishing in their **free zones**, spaces where taxes and tariffs, as well as the cost of labor, are lower due to exemption from state regulation (Ong, 2006). The finished electronics are then sent to the US to be sold in high-value markets, clearing a lot more profit than if they had finished them in Japan. The profit on bananas is comparably low, so filling a container with bananas headed to the US, when it could be filled with electronics, does not make sense. But filling an empty container headed back to Japan (where they finish their own electronics for domestic consumption), also known as back haul, makes a lot of sense and is a supply chain relationship that banana producers have used to get their products into higher-value markets. This whole system makes sense when all the relationships are considered and when power in the form of private capital is assumed to take priority over other forms of welfare, such as sustainability, equality or health.

Provided in this chapter is a brief summary of how geographers think about the world and concepts and theories that we use to understand and analyze problems and empirical observations. Following that is a discussion of the way power in the form of food system governance, neocolonial policy and uneven development influences human health and environment, sustainability, and inequality for differently situated people.

DOI: 10.4324/9781003159438-3

FIGURE 2.1 More than a hundred million bananas move through the San Diego port each month (https://commons.wikimedia.org/wiki/File:Dole_Fresh_Fruit_at_Tenth_Avenue_Marine_Terminal_(1).jpg)

THINKING GEOGRAPHICALLY

Human geographers study how human populations are distributed and organized spatially and the consequences of that distribution. They also study the lived spaces, the meanings of places, and how we come to know the world. Physical geographers study how the earth is created, while human geographers study how humans modify and change the planet to meet their needs. Techniques geographers develop and use, such as maps, geospatial data, satellite-image analysis, and an especially powerful set of tools for analysis called "Geographic Information Systems" or "Science," or "GIS." Both human and physical geographers use techniques to understand how the world is created and modified. The holistic nature of geography is both a strength and a weakness. Geography's strength comes from its ability to identify interrelationships that are not normally noticed in narrowly defined fields of knowledge. For example, genetically modified seeds are produced by physical science, but their human impacts must be studied with social science. Geographers study both at the same time. Another example is the study of infrastructure development in urban areas. To understand where to lay sewer lines or other hydrological infrastructure, urban planners must understand groundwater as well as patterns of urban structure. The most obvious weakness associated with the geographic approach is related to the fact that holistic understanding is often overly broad or global in its understanding and could miss important details of cause and effect. Human geographers use a handful of conceptual tools to understand and explain patterns and interrelationships in the world. In what follows, three sets of paired concepts are described.

Space and Scale

Space is a basic building block for thinking geographically. Space has traditionally been thought of as a container or something that holds certain activities or people. It has readily identified bounds or edges. Doreen Massey (2005) conceptualized space as open ended and infused with power and co-created with other spaces. Given that, geographers think about space in three different ways: absolute, relative, and cognitive. Let's use a classroom to demonstrate. The first way to think about space is in terms of its dimensions or properties, such as how big it is or seating capacity. The second is a comparative sense, that is, this classroom is larger or smaller than others. The last is the emotional or affective realm, which is the feelings people associate with the space, which may be boredom or excitement, anticipation or dread. People associate certain things with spaces, and those feelings or attachments become part of what makes it a place, as described later in this chapter. These three examples also illustrate the concept of scale.

The classroom is but one space within one census tract within a city. This classroom is thus situated within a nested hierarchy, called **scale**, of a campus, city, state, parish, county, province and country. While the classroom, the city and the province are all part of the same space of country, bounding it differently allows for measuring or comparing different spaces. Scale is therefore a way to partition space, and this is often done in a hierarchical way so that patterns of interrelationships can be revealed. Scale, as discussed by Marston and Knox (2007), is divided into four realms – the global, the world region, nation-states, and human settlements. Scale, as a concept, is not the same as mathematical scale, which is used to indicate how much space is captured on a map and the ratio of representational space, the map, to what is represented, the world.

Distance and Networks

Like space, distance has three dimensions: absolute, relative and cognitive. Absolute distance is simply the distance in units (miles, kilometers) between things, and relative distance would be the result of a comparison between two differences. Cognitive distance has to do with perception and feelings and is often captured by the idea of **friction of distance**. Friction of distance suggests that there are costs and tensions associated with absolute differences in location (i.e., the expense of commuting between home and work). In a more cognitive realm, this can relate to something called **social distance**, which is the way in which certain groups try to separate themselves from other groups. The largest social distance would be by exclusion from a country through legal or extra-legal means, while the shortest social distance would be intermarriage. Historically, people in different racial groups were forbidden to intermarry, and anti-miscegenation laws were only struck down in 1967 in the US. Through this legal mechanism, distance between social groups was maintained. In another example, according to Jackson (2006), philanthropic giving often fosters caring for distant strangers whose plights often receive media attention, and thus bridge social distance, over those in our neighborhoods who do not.

Distance is often thought of as what is between two nodes in a network, which is also known as **topological space**. Topological space concerns itself with the nature and degree of connection between places, regions, landscapes, locations and so on. The example map shown in Figure 2.2 is the subway system of Beijing, China, and shows the way points in the city are connected. It is not true to scale or the geography of the city – it simply shows how points are connected in what are known as "vectors" or linear space. Recall from the "Laws of Geography" that distance is often trumped by connectivity (i.e., things that are closer may be less accessible because they are not well connected). The map in Figure 2.2 shows, in the lower right corner, how the last station on the Yizhuang Line is quite far "topologically" from the last station on the Batong Line, but they are actually quite close in absolute terms, as is quite often the case, the farther something is out from the center of

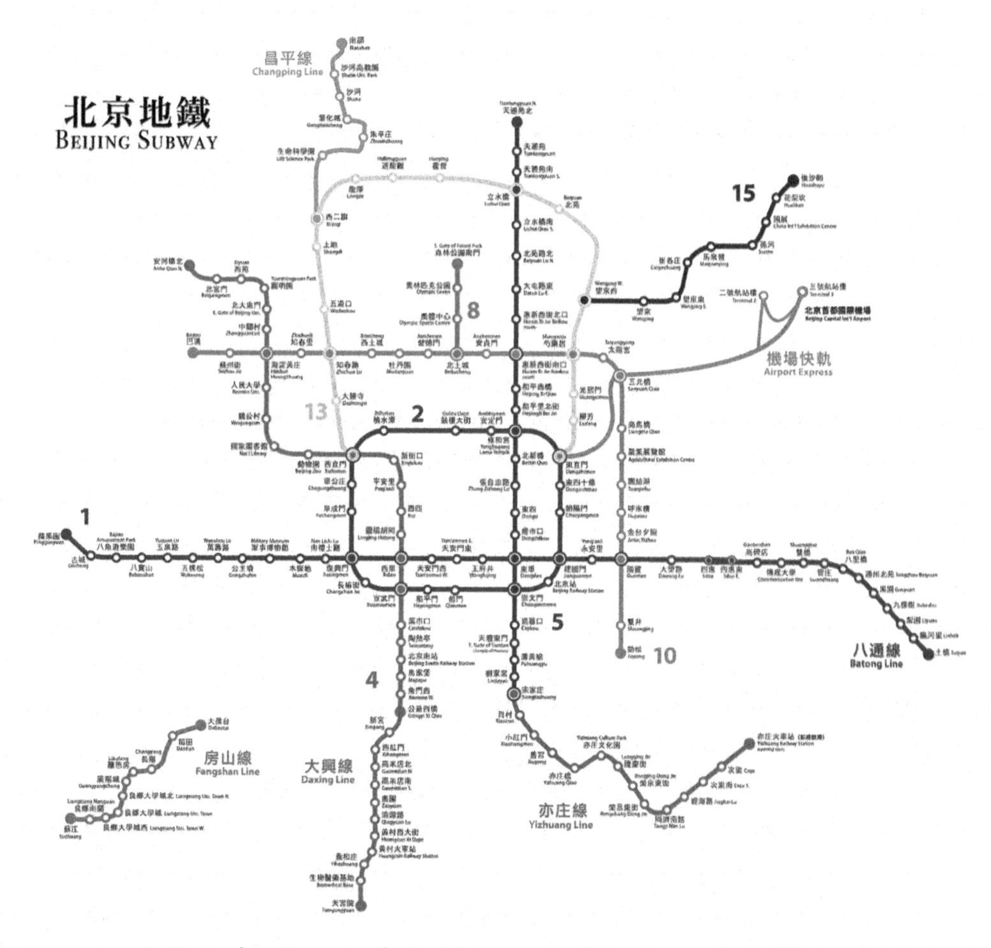

FIGURE 2.2 Beijing subway system (https://commons.wikimedia.org/wiki/File:Beijing_Subway_in_Hant_in_2011.png)

a network. Accessibility is also a function of social, cultural or economic factors. One may be better "connected" to others through cultural capital (i.e., intermarriage across race) or through economic power (i.e., one can use money to gain access to legislators).

Place and Region

Places are central to the processes of the scale of human settlements, and the concept of place helps geographers understand the meaning of its sub-realms: home, community, body. Place is simply defined as a *space with meaning*, although it is always and already much more than that. Thinking about place as a *cognitive space* however, lets us focus on how meanings, feelings and associations become attached to spaces through interactions between people and groups (Cresswell, 2014). Places help people understand who they are and how the world works. They provide opportunities for well-being and a context for understanding and contesting social norms. A place could be a home, a neighborhood, an institution such as a university, or a city, and as such, a place is bound by spatial imaginaries. However, geographers such as Massey (2005) assert that flows between places shape them as much as the way they are bounded by imagined limits.

A **region** is thought to be a grouping of places that are **socially constructed**, which means that their meanings are assigned through social interaction. Geographers divide regions into two categories: formal and functional. *Formal regions* share some kind of internal homogeneity based on shared characteristics, such as language or religion. *Functional regions* are internally diverse but are designed to work together toward some political or economic goal. The definitions of regions are often highly contested and reveal that commonsense knowledge is often highly contextual. For example, what we commonly think of as the "Middle East" varies widely in terms of its geographic bounds and what countries and territories are included in it or not. Throughout the book, reference is made to collections of countries as regions to indicate that countries work together to produce a commodity, such as shrimp in Southeast Asia, which includes countries that are very diverse: Thailand, Malaysia, Papua New Guinea, Indonesia and Myanmar. Or we will talk about a group of countries that have a shared experience of colonialism and appropriation of food producing resources, such as South Asia, which includes India, Pakistan, Bangladesh and Sri Lanka.

THINKING SPATIALLY AND RELATIONALLY

Thinking Spatially: The Dead Zone

So, let's put these concepts into action with a discussion of the dead zone in the Mississippi River delta in the Gulf of Mexico. The delta is a product of both human and natural systems. Corn is a crop heavily subsidized with federal money, which encourages farmers in the American Midwest region to grow a lot of it. Corn requires heavy inputs of nitrogen **fertilizer**, which is any material, natural or synthetic, used to supply nutrients to plants, to meet market-based yield standards. Typically, a farmer has to produce hundreds of bushels an acre to make a living and to do that requires massive inputs of nutrients from outside the system. Corn is planted intensively along the Mississippi River and its tributaries. The nitrogen fertilizer finds its way to bodies of water via the mechanism of **runoff**, or the movement of water from land into bodies of water.

When water meets land, it slows down and deposits its sediment and runoff into the shallow, warm water of a fertile and productive river delta. Once there, the fertilizer stimulates the overproduction of marine organisms, which use up the available oxygen in the water when they die and decompose. The deeper water becomes a dead zone, where marine life can no longer live, which has a **trophic** or cascading effect on the food web, killing off entire marine ecosystems. The dead zone is not a product of the natural environment or natural systems alone. The fertilizer that is added to the environmental system is human-produced and distributed in specific political-economic ways. The goal of geographic thinking is to *reveal patterns of relationships* in a wide spectrum of natural, social, economic, political and cultural phenomena, which demonstrates how and why geography matters to everyday life. Figure 2.3 shows low oxygen levels on the Gulf Coast (red indicates lowest readings) that starve marine life of necessary oxygen.

Geographers also demonstrate *how places and processes are related* (i.e., how the American Midwest is related to the Mississippi delta via the socio-natural system of agriculture). The emphasis on **interdependence**, interrelationships and connection in geography means

FIGURE 2.3 Mississippi River "dead zone" (NASA/Goddard Space Flight Center Scientific Visualization Studio)

that geographers can often solve complex problems because they have a greater understanding of the interconnections between places and processes. In the case of the delta's dead zone, geographers might recommend a change in the commodity corn subsidies as a more direct solution to the problem than encouraging farmers to use less fertilizer or attempts to remediate the environment in a delta. Geographers also draw attention to the importance that **distance and connectivity** play in relationships. In a general sense, "near" things are more related than distant things. This definition of "near" must, however, include consideration of how **networks** create close connections. While corn production may not be physically near the delta, it is connected very well via the Mississippi River. Geographers also understand that one process may be experienced at multiple places and scales simultaneously through the partitioning of space. One example includes how the Soviet invasion of Afghanistan in the 1980s drove bankruptcy in the farming communities of the US. The causal link was an embargo on grain shipments to the USSR, which at the time was one of the largest buyers of American commodities. Links between places that are not obvious on the surface can be established by following the ways in which they are connected socially, economically or politically.

In another example, a shrimper fishing in the Gulf of Mexico may experience losses due to the dead zone, while a farmer in Iowa will receive a big subsidy check for a bumper crop made possible by fertilizers. As a result, the economy may suffer in one place, while the economy in another will flourish. And an entire region may relationally decline, relative to the standard of living of people in another. The **nation-state** might benefit from the circulation of corn in the **global economy**, while an entire community might languish through loss of livelihoods. Thus, what happens in one place, region or scale is related to what happens in another place, often in an inverse relationship. The relative affluence of farmers in Iowa due to subsidized corn production is related to the financial decline of shrimpers in the delta. Thus, places are interdependent, and interconnected, and often shaped by relationships of **unevenness**.

Thinking Relationally: The Intersectional Zone

Categories take on meaning through their relationship to other categories. Often referred to as **binary thinking**, this is a cognitive process in which one category, such as "rural" becomes defined by what it is not: "urban." Feminist scholars have long critiqued this way of thinking, because it is often subconsciously value laden, meaning that things on one side of the binary are seen as implicitly more desirable than the other and are often connected to other desirable ideas or attributes. For example, concepts arising from Western European Enlightenment, such as rationality and reason, were associated with men, while women were viewed as irrational and emotional. Scholars of feminism and sexual orientation and gender diversity have unpacked how categories take on meaning to expose the values and ideologies that support femininity and masculinity as binary categories (McDowell, 1993). Exposing this pattern creates an opening to think about gender identities as fluid and overlapping, rather than fixed and static categories. It also raises awareness about intersectional identity formation, or when identity takes on multiple meanings through the accretion of many kinds of difference and oppressions.

Intersectionality views subjects as produced at the intersection of many forms of identity: race, class, gender, sexuality, ethnicity, nationality, ability, religion and so on (Taylor, 2017). Intersectionality as a concept arose from critiques of feminism by Black and queer women of color in the US. Black feminists extended this critique and argued that the embodied experience of raced, classed and gendered individuals is shaped by a "matrix of domination" (Hill-Collins, 2002, p. 18) that contributes to the dehumanization of entire groups of people (McKittrick, 2006). Thus, economic inequality is not a single form of oppression that all people experience in exactly the same way. Rather this form of oppression is a social relationship defined by one's ability (or lack thereof) to perpetuate and benefit from certain kinds of privileged positions due to race, gender, sexuality, citizenship status, ability and so on, that vary across space and time. In the case of food and agriculture, the experience of hunger or malnutrition is different for different groups of people through uneven socio-economic relationships. These uneven, often exploitative, relationships are shaped by global capitalism that furthers the accumulation of capital to certain individuals (i.e., billionaires) in the private sector.

To return to the example of the dead zone, the shrimper in Mississippi is likely working class, male and Black, or possibly a working-class White woman (Wilson, 2005), running a small-scale family business selling wild shrimp caught in public waters in local or regional markets. The Midwestern farmer is very likely White and male, running a technically small business, but large in comparison to shrimpers, with close to a thousand acres of private land in production, and produces a product destined for global markets and heavily subsidized by tax money. The industry behind corn growing is composed of some of the largest and most profitable corporations in the world, and its lobbyists are powerful. The shrimp industry has no such equivalent. Thus, power is integral to the story that plays out here and is essential for explaining why the relatively powerless shrimpers endure the loss of livelihood and polluted environments. Geographers can reveal patterns of **interconnection** and relationship and the impacts they have on individual people and social groups in intersectional ways, but this requires an understanding of power to understand *why* they come to be and how they persist.

Geography, Power and Food

Power is generally understood as the ability to have influence or to shape the actions of individuals or of groups, such as governments or institutions such as universities. Power comes in two distinct varieties: the first is the kind that inspires action through sharing a worldview or an idea about the right way to do things. This kind of power is ubiquitous in daily life and is as mundane as how a person might cook food or choose what food to eat or as significant as foreign policy decisions governing the import of foods from some countries or food safety regulations banning the production of food in some places. These are laws that are generally accepted as working toward the collective best interest, and as long as people share that vision for the world, people go along with it. Another kind of power is coercive, which uses violence to obtain its ends, or bears the threat of violence, to obtain them. Violence can include bodily harm or loss of freedoms or rights. Sovereign nation-states have a monopoly on coercive power and may inflict violence on their citizens with impunity. The death penalty and genocide are notable examples. The extermination of bison to quell Indigenous people in North America and the use of slavery (contemporary and historical) to produce food are two examples of coercive power in the food system.

Governance is the act of using power to shape policy and laws, usually by institutions such as the state in cooperation with other actors and sectors. The diagram in Figure 2.4 shows the state, the market and civil society as three forces that act on the food system to shape what food is available, to whom, where and how and under what conditions. In a general sense, the state, usually as the federal government of an independent country, acts as though food is in the interest of the general public or a *public good* that everyone is entitled to have. This could take the form of providing adequate nutrition for people through food distribution programs or through ensuring the safety of food that people consume through regulation. The market, composed of corporations, private enterprises, small businesses and stockholders, acts as though food is a *private good* to be bought and sold to facilitate the production of profit for the benefit of some, but not all. Civil society is a large, diffuse collection of individuals, groups and institutions that together comprise a realm of social life characterized by reciprocity and volunteerism and that view food as a *common good*, or something to be shared and not sold or regulated. Community gardens or the sharing of food from backyard gardens through informal networks would be examples of food as a common good in civil society.

Each sector has two aspects to it; the aspect on the outer edge of the inner triangle (informal, for profit, public) characterizes that sector, while the aspect on the inner edge of the triangle has an aspect that overlaps or intersects with other sectors. For example, civil society shares food as a common good through a community garden in an unused piece of land in an urban neighborhood, which would be an informal example. Another

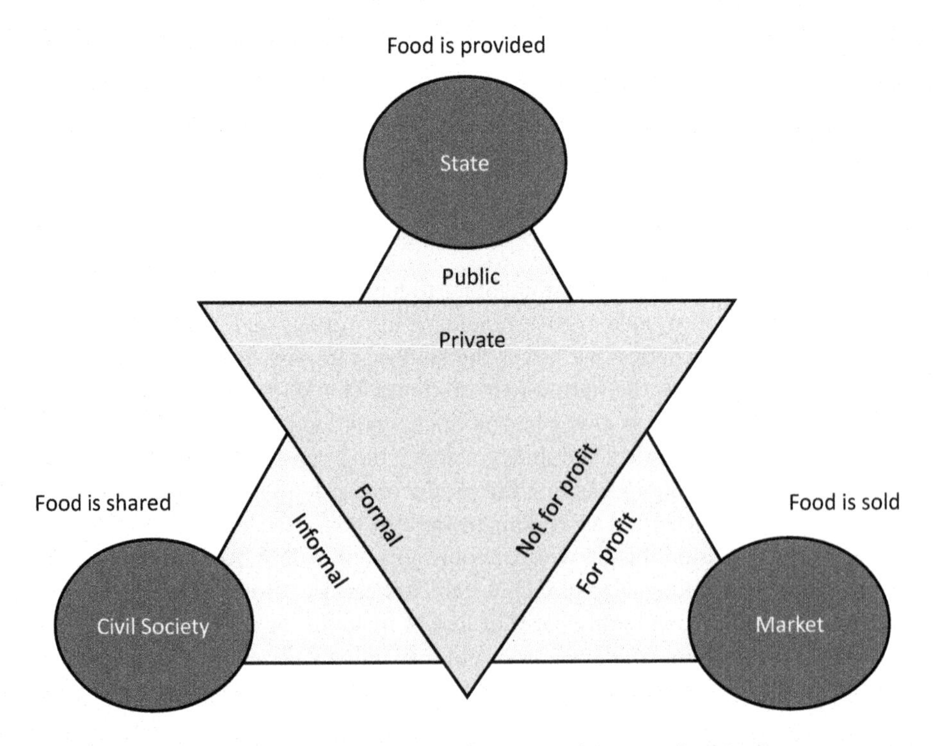

FIGURE 2.4 Food system governance (adapted from De Schutter et al., 2018)

community garden may give food away but be sponsored by grants from a corporation and regulated by state laws in ways that are formalized through regulation. It may also be a 501c3, or a not-for-profit corporation, which would entitle it to have bylaws and stakeholders who would have the right to keep it private, meaning it would have the right to exclude some people from taking the food, and it could accumulate assets for private use. Many corporations have not-for-profit extensions that provide capital for these kinds of experiments, and the state has laws that regulate them. Each sector may use its power to obtain its ends: the state through its political power to shape policy and access to food and corporations to set prices and influence labor markets and civil society to shape collective ideas about what is right, just and fair.

Geography and power come together in the food system with relational and intersectional effects in a multitude of ways. For Ruth Wilson Gilmore (2002), space and power are twin forces acting to subject groups to what she calls "premature death," or fatalities caused by the unjust use of power against particular bodies over time and in specific spaces. She writes,

> **Racism** is a practice of abstraction, a death dealing displacement of difference into hierarchies that organize relations within and between the planet's sovereign political territories. Racism functions as a limiting force that pushes disproportionate costs of participating in an increasingly monetized and profit-driven world onto those who, due to the frictions of political distance, cannot reach the variable levers of power that might relieve them of those costs.
>
> (Gilmore, 2002, 16)

"**Environmental racism**," according to Bullard (1993), "refers to any policy, practice or directive that differentially affects or disadvantages (whether intended or unintended) individuals, groups, or communities based on race or color" (1037). **Environmental justice** advocates work to raise awareness or mitigate the negative impacts of pollution that are disproportionately found in communities of color.

The delta is a place polluted with not only fertilizers but also **pesticides** (which are any substances used to kill organisms harmful to plants and/or animals and include insecticides, fungicides and herbicides), which are sprayed alongside and with similar volumes as fertilizers on corn and also soybeans. Conventional agriculture in the case of the Mississippi delta wields downstream death through pesticides to those living in the delta, as well as the flat plains of the areas upstream from the delta. Pesticides are known endocrine disruptors, carcinogens and neurotoxins, leading to metabolic disruptions such as obesity and type 2 diabetes (Guthman, 2011), cancer and neurological disorders after long-term, chronic exposures, such as those experienced by people "living downstream" (Steingraber, 2010). It should come as no surprise then that the rates of those diseases, particularly type 2 diabetes, in the delta exceed national averages.

It should also come as no surprise that the rates of people living with such diseases upstream are far lower than the national average. They, largely White, male and middle class, are also the ones who are directly benefiting from the application of pesticides (and fertilizers), and their comparative standard of living and life expectancy are far higher than those who live downstream, who are disproportionately Black, female and poor or working class. Gilmore (2002) writes, "if justice is embodied, it is then therefore always spatial, which is to say part of a process of making a place" (16). The making of the place of the Midwest as a place of prosperity through agriculture comes at the expense of the health of the environment as well as the people who live downstream, in a relational and unjust displacement of the goods and "bads" of agriculture, as advocates for **food justice** have argued. This is a socially mediated process that plays out in the food system across multiple scales and regions and connects people across distance in uneven and inequitable ways. This is a process driven through the prioritization of profit on two powerful arms of the food system: the state through its subsidies of particular crops and the market in the way that it supports the accumulation of capital in the private sector. Capitalism, as a system that generates inequality and was built on the architecture of colonialism and imperialism, rewards one group at the expense of another in what Harvey (2003) calls **accumulation by dispossession**. This process is always spatial and always social. As such, it is a system that is created by people and can be changed, as will be demonstrated through the remainder of this book.

SUMMARY

This chapter began with an example that underscored the primacy of profit to the architecture of the food system and how those pathways were laid through the power of colonialism. The power of capitalism now drives global production and distribution processes for all kinds of goods along the supply chains that imperialism forged. Capitalism and the pursuit of profit drive the distribution of foods as well as the goods and "bads" of agriculture. Geography is a discipline that uses conceptual tools such as space and distance to explain how and why certain power relationships appear in the world and what societies can do about them. Thinking relationally and intersectionally about the world allows for an analysis that examines injustice and unevenness in agricultural development that undermine sustainability, the health of humans and the environment and generate the perpetuation of inequality. As a social system created by people, it can be changed.

SUPPLEMENTAL MATERIALS

Key Terms and Concepts: accumulation by dispossession, binary thinking, connectivity, distance, environmental justice, environmental racism, fertilizers, food justice, free zones, global economy, governance, interconnection, interdependence, intersectionality, nation-states, networks, pesticides, place, power, racism, region, runoff, scale, socially constructed, space, topological space, trophic, unevenness

Explore More: Peter Menzel published a book of photos of people and the food they eat for a week, called *Hungry Planet*. Explore his photos here under the Hungry Planet tab to understand the differences that space and place make. www.menzelphoto.com/index

Recommended Viewing: *Living Downstream* (2010), *Come Hell or High Water: The Battle for Turkey Creek* (2014), *Urban Roots* (2011)

Discussion Prompt: Everyone uses geographic information in the form of cognitive maps to navigate the world to obtain food. In 50–100 words, list three ways you used geographic information today or in the past week as it relates to food procurement. Try to imagine ways that go beyond using GPS or Google Maps. Think about how your brain tells you that you belong or do not belong in certain spaces, or what kinds of spaces feel familiar to you because of your memories and associations with them and how this is a relational and intersectional process. Be sure to correctly use at least *one concept* from this chapter in your discussion post and to write it in **bold** font.

Recipe: A low-country shrimp boil is a traditional preparation of shrimp in the Gulf of Mexico and Atlantic Coast. Look for farmed or wild Northern Shrimp from Canada or the Western US given that Gulf shrimp are overfished. www.foodnetwork.com/recipes/trisha-yearwood/low-country-boil-recipe-2112348

REFERENCES

Bullard, R. D. (1993). Environmental racism and invisible communities. *West Virginia Law Review*, *96*, 1037.

Collins, P. H. (2002). *Black feminist thought: Knowledge, consciousness, and the politics of empowerment*. Routledge.

Cresswell, T. (2014). *Place: An introduction*. John Wiley & Sons.

De Schutter, O., Mattei, U., Vivero-Pol, J. L., & Ferrando, T. (2018). Food as commons: Towards a new relationship between the public, the civic and the private. In *Routledge handbook of food as a commons*. Taylor & Francis.

Gilmore, R. W. (2002). Fatal couplings of power and difference: Notes on racism and geography. *The Professional Geographer*, *54*(1), 15–24.

Guthman, J. (2011). *Weighing in: Obesity, food justice, and the limits of capitalism* (Vol. 32). University of California Press.

Harvey, D. (2003). *The new imperialism*. Oxford University Press.

Jackson, P. (2006). Thinking geographically. *Geography*, *91*(3), 199–204.

Marston, S., & Knox, P. (2007). Places and regions in global context. In *Human geography*. Pearson Prentice Hall.

Massey, D., & Massey, D. B. (2005). *For space*. Sage.

McDowell, L. (1993). Space, place and gender relations: Part II. Identity, difference, feminist geometries and geographies. *Progress in Human Geography*, *17*(3), 305–318.

McKittrick, K. (2006). *Demonic grounds: Black women and the cartographies of struggle*. University of Minnesota Press.

Ong, A. (2006). *Neoliberalism as exception: Mutations in citizenship and sovereignty*. Duke University Press.

Steingraber, S. (2010). *Living downstream: An ecologist's personal investigation of cancer and the environment*. Da Capo Press.

Taylor, K. Y. (Ed.). (2017). *How we get free: Black feminism and the Combahee River Collective*. Haymarket Books.

Wilson, D. (2005). *An unreasonable woman: A true story of shrimpers, politicos, polluters, and the fight for Seadrift, Texas*. Chelsea Green Publishing.

PART 2

Food for Thought

CHAPTER 3

The Origins of Food

INTRODUCTION

Colonizers arriving in the Americas encountered what they considered a "tabula rasa" or a blank slate of unoccupied land free for the taking. Under the white supremacist gaze of the colonizers, who believed only Europeans knew how to do agriculture, forests, parklands and grasslands were viewed as uncultivated. Not recognizing the oaks of the mid-Atlantic or the rain forests of the Amazon or the grasslands of Australia as cultivated landscapes also served the white supremacist and genocidal projects of European colonizers, who considered agriculture an important hallmark of civilization that would imply a prior land claim. Thanks to the efforts of Indigenous scholars and activists, including significant research into the ethnobotany of pre-Columbian forests, land that had been cultivated for centuries by Indigenous people has finally come to be seen as agriculture. The Brazilian Amazon (Mann, 2005), the Australian Outback (Pascoe, 2014) and many North American landscapes (Kimmerer, 2013) are now seen as cradles of domestication and cultivation of important plants, albeit in ways that were unrecognizable by European invaders.

The oak savannah pictured in Figure 3.1 is an example of a landscape shaped through Indigenous methods that was seen as "naturally occurring" by European colonizers. The persistence of forest food species is an important source of resilience for Indigenous communities under ongoing conditions of colonization, but they face new threats due to development, climate change and the spatial logic of capitalism. Agriculture began as a way to minimize the labor required to acquire enough calories for populations to feed themselves, thereby freeing labor for specialized tasks such as weapons-making or trade. The earliest forms of agriculture are thought to be simple forms of **pre-domestication cultivation** or the selective sowing of seeds or the culling of certain animals in particular places in a hunter-gatherer's environment. This practice of species selection has since grown into a trillion-dollar industry utilizing the latest technologies in artificial intelligence and cellular manipulation in the form of genetic modification and laboratory-grown products. This chapter examines the origins of agriculture as we know them and traces agricultural change from the first domestication to the latest applications of technological innovations in food production.

DOI: 10.4324/9781003159438-5

FIGURE 3.1 Oak savannah in Kentucky (https://commons.wikimedia.org/wiki/File:Bald_top_oak_savannah.jpg)

KEY THEMES AND IDEAS

For most of prehistory, humans obtained food through the parallel and complementary practices of **hunting and gathering**. In this world of precarity, tubers, berries and seeds were reliable sources of carbohydrates that were complemented by the acquisition of protein through hunting, which was not always guaranteed. The use of the fire was also central to the development of the human brain, as fire made food easier to digest. Therefore, energy previously used by the gut could be freed up for use by the brain to develop new tools and languages (Wrangham, 2009). Until recently, it was thought that men were hunters and women were gatherers, based on tribes such as the San in Africa, who maintain such segregation, but new DNA testing of skeletons in the Andes and central Asia suggests that women were also hunters (Haas et al., 2020). Hunter-gatherers also tended to be nomadic, following the availability of game and foraging during the changing seasons. As such, land claims were fluid and often overlapping or shared. At some point in various places in the world, humans figured it was easier to keep the foraged foods and game animals closer to home. They made permanent claims to land, built enclosures and began domesticating plants to improve their productivity. Drivers of this change remain unclear, but perhaps climate change, human population pressure or cultural practices were influences.

In what follows, a basic history of agriculture is outlined, framed around the three major "revolutions." The revolutions mark profound changes in the production and distribution of food, and along with them the application of technology and new social relationships.

These revolutions coincided with, drove and benefited from massive global shifts in political economy. Given this, scholars have also suggested that modern food production and distribution fall under varying forms of governance over time and across space. Shifts in power relations around control of food quality, quantity and availability signal **regime** change, or changes in what the food system provides, to whom and how. Thus, agricultural revolutions and food regimes have some overlapping and complementary characteristics, which are described in the next section. In addition, they occur in roughly similar time frames, which is indicated in Figure 3.5.

The First Agricultural Revolution

The conventional history of agriculture begins with the domestication of barley, wheat, sheep, pigs and goats around 12,000 years ago in the so-called Fertile Crescent in the environs of present-day Iraq, Syria, Israel/Palestine and Turkey. According to conventional histories, peoples in other regions independently developed agricultural innovations, at two different time periods in the early and middle Holocene. They include regions of North and South America; West and East Africa; South, East and Southeast Asia; and Oceania. History tells the stories of the colonizer, and so much of the knowledge of these agricultural systems has been lost to multiple waves of imperialism in these places. What we do know from archeological records and from the accounts of Indigenous people, is that agriculture emerged as an innovation central to societies up to 65,000 years before present in multiple locations (Pascoe, 2014).

Figure 3.2 indicates the place and time period for the emergence of agricultural innovations in different world regions. Each major inhabited world region had an independent hearth for domestication in roughly the same time frame of the early Holocene (marked by a circle) and again in the middle Holocene (marked by a rectangle). These dates are derived from carbon dating of artifacts found through archeological research. Other sites of domestication, marked by triangles, are **biogeographically inferred** or assumed to be such due to the presence of cultivars and source plant genetics of modern varieties. For example, African rice was domesticated in what is now Mali about 3,000 years ago. It is closely related to other varieties of rice domesticated in Asia but is significantly divergent genetically, suggesting an independent origin. It is not widely cultivated outside of Africa but is of interest to scientists for its unique resistance to environmental stressors, including salt tolerance.

In a general sense, agriculture has followed a pattern of domestication and innovation over the past 12,000 years, with variations on that theme: new plants and animals considered suitable for agriculture have slowly been incorporated into diets and cultivation systems. New techniques and methods of producing, distributing and preparing food for people to consume have also been developed and incorporated into the global food system over time. The **first agricultural revolution** is thought to have occurred 12,000 years ago in the early Holocene when hunter-gatherers considered the possibility that keeping plants and animals closer to their settlements represented a savings of energy. Bruce Pascoe writes about early Australian agriculture in his book *Dark Emu* (2014). Using records of early explorers to the Australian continent, he pieces together evidence that aboriginal Australians domesticated plants and kept animals in semi-domesticated situations. Kimmerer

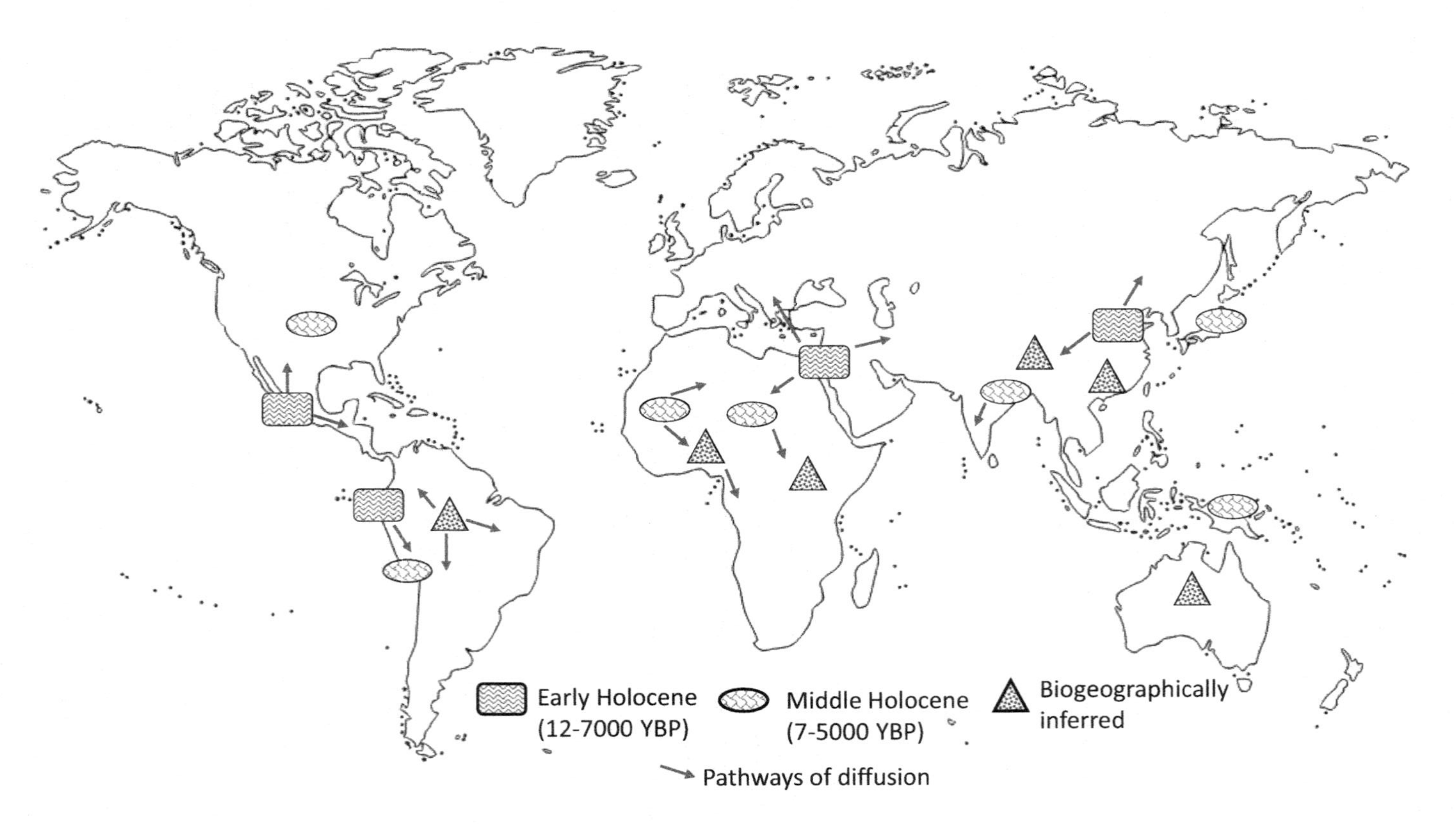

FIGURE 3.2 Origins of domestication (adapted from Larson et al., 2014)

(2013), in her book *Braiding Sweetgrass*, also documents how early pre-domestication in North America consisted of small interventions in the natural habitats and life cycles of plants and animals that Indigenous people depended on for food, and this continues in many communities today.

Domestication is generally thought to be a product of selective adaptation to place and region, and perhaps also changing environmental conditions (Larson et al., 2014). Human preferences for certain foods, such as ease of harvesting and edibility, certainly played a role in which plants and animals were domesticated. Characteristics of domesticated plants include a dependence on human intervention for seed dispersal, an increase in seed size, a reduction in bitter chemicals, and more predictable germination. The earliest crops to be domesticated were wheat and rice, in the present-day region of the Middle East, and these were for centuries the only known domesticates. Although it is still unclear what drove people to domesticate plants, it is clear that prehistoric humans modified their environments to make them more favorable to cultivating plants (e.g., the enriched soils found in Amazonia). The sharing of cultural practices within and between groups is also a significant factor in the perpetuation and spread of agriculture and is no longer underestimated in domestication studies. Terracing, canals and even castration of animals are well documented in archeological records, and such cultural practices point to the importance of traditional ecological knowledge in maintaining and extending into the present the varieties of plants and animals that we consume.

Dogs were the first animals to be domesticated, in the late Pleistocene, well before plants anywhere, although an exact date is uncertain. In some places (Africa, Arabia, and India), animals were domesticated before plants, and in others (South and Central America), the opposite is true. Genetic changes found in domesticated animals, known as **domestication syndrome**, include changes in hormones that render them more docile and make reproductive cycles more predictable, different coat colors and other changes to body size and proportions. It is clear from archeological records that the same animals were often domesticated independently in different places (i.e., pigs in South Asia and the Middle East). Humans acted in multiple ways to bring animals into closer and more permanent relationships with humans (and for more reasons than food). They managed the mobility and numbers of livestock, selected for certain traits (docility) and changed the environment to encourage livestock to remain close to human settlements or in circumscribed locations. This could include the cooperative use of fire to maintain grazing lands, cultivation of crops resistant to overgrazing and the use of natural barriers, such as rivers and canyons.

What developed from these early interventions was something that we now call **subsistence agriculture**. Still practiced by people the world over in rural communities, subsistence food production is composed of diversified plant and animal systems, which are often integrated into each other through interdependence. For example, the dung of domesticated animals is used as fertilizer for the crops that both humans and animals eat. The animals provide protein through meat, milk or eggs that complements the carbohydrates from crops such as wheat, oats or rice. Most subsistence agriculture was thought to be rain-fed, but new evidence suggests sophisticated wells and canals were used to channel water in many early agricultural societies. The products of historic subsistence were dedicated to feeding the community and saving seeds for the next season. Surpluses were stored for future use or planting or traded for other goods. In this way, subsistence agriculture differs from commercial agriculture; the food is grown primarily to nourish families and communities rather than for profit. Consequently, few inputs are used, and energy loops are often closed to preserve resources. For example, in addition to animal manures, **green manures** – those plants that deliver nitrogen to the soil, such as soybeans or peas – were widely used to augment soil fertility and to provide plant-based protein to the diet. All subsistence agriculture then and today relies on open-pollinated seeds, which are **traditional cultivars** (seed varieties developed in local areas by farmers) that breed true and can be saved for future planting. Early farmers would have saved seeds for planting the next year based on desirable traits such as taste, hardiness or productivity. The power relationships that shape subsistence agriculture are often shaped generationally, across the community and within the household. Many researchers have documented the dominance of elder males in governing such systems, but new research indicates that the women who do the work of agriculture often had (and still have) much decision-making power about what to plant and how to distribute the products (see, e.g., Twyman et al., 2015).

FOCUS SECTION: SEEDS

Agriculture changed very little between its earliest periods and the modern era. Michael Pollan likes to say that the way we eat has changed more in the last 50 years than in the last 10,000, and that is nowhere more evident than in our seeds. There are three different kinds of seeds in use in our food system. The first is what is called **open-pollinated**, or heirloom seeds that are the result of centuries of subsistence agriculture using **landraces**, or naturally occurring seeds in ecological niches. These are seeds selected and bred in particular places over centuries that fertilize naturally, with insects or wind, from pollen of the same variety. They breed true over generations but may contain genetic diversity that they do not necessarily express as traits. The second type, developed in the Green Revolution in the middle of the 20th century, is **hybrids** that are bred for particular traits. They are created in laboratory settings by artificial inbreeding of different varieties of the same species. These seeds cannot be used for successive generations because they exhibit trait variability in the second generation, much like a donkey-horse cross results in a mule. The last kind of seeds are **genetically modified**, whose genomes are altered by the insertion of genetic material from a different species. Bt corn is widely used in agriculture today, and it was created by splicing bacterial DNA into corn to develop resistance to pests.

Seneca Pink Lady corn, shown in Figure 3.3, is an Indigenous developed variety of pink- and red-colored corn (as opposed to the more common white, yellow and blue) that is grown for cornmeal by a First Nations tribe in New York. The remnants of an ancient variety preserved in spite of genocide and dispossession were given to the White Earth Anishinaabeg in Minnesota. The White Earth Land Recovery Project grew the remnants

FIGURE 3.3 Seneca Pink Lady corn

out as their own special corn variety, which they grind into cornmeal and sell to fund their operations.

Recall from Chapter 1 that the application of technology to agricultural inputs, such as seeds, results in increased agricultural productivity and an increased food supply. This accompanies, generally, a growth in population, given the new calories available to the population. It also results in migration away from rural areas because technology often decreases the need for labor and allows people to do something else with their time besides work in the farm fields. This relationship was one that had a mathematician named Malthus worried in the 1800s. He saw human population growing at an exponential rate and knew that, at least theoretically, food availability could only be increased arithmetically by cultivating more land. His dire predictions of widespread famine were wrong, partly because he was looking at one part of a complex demographic transition and also because he could not foresee the technological innovations in agricultural systems that allow for more food to be produced on the same land. Recall from Chapter 1 and Figure 1.2 that applications of technology to agriculture increase food supplies, concentrate farmland and trigger migration, urbanization and population growth.

The Second Agricultural Revolution

Another wave of innovations in agriculture composes what is thought to be the **second agricultural revolution**. This began in the 1600s in Europe where efforts were focused on innovating at a genetic level through plant and animal breeding. Also known as **scientific agriculture**, this effort to improve plant varieties and animal breeds dovetailed with the European project of expansion and acquisition of territory for capitalism. The scientific revolution used Mendelian genetics to understand how plant breeding worked and how to introduce new traits into plants and livestock. In the famous example shown in Figure 3.4, Belgian blue cattle were bred for extreme muscling and therefore greater meat production in one animal. This time period was also the beginning of what Friedmann (1987) identifies as the **first food regime**, which lasted until World War I, during

FIGURE 3.4 Belgian Blue bull, exhibiting muscling through selective breeding (by agriflanders – originally posted to Flickr as Kamp-Bambino vd ijzer copy, CC BY 2.0, https://commons.wikimedia.org/w/index.php?curid=6257487)

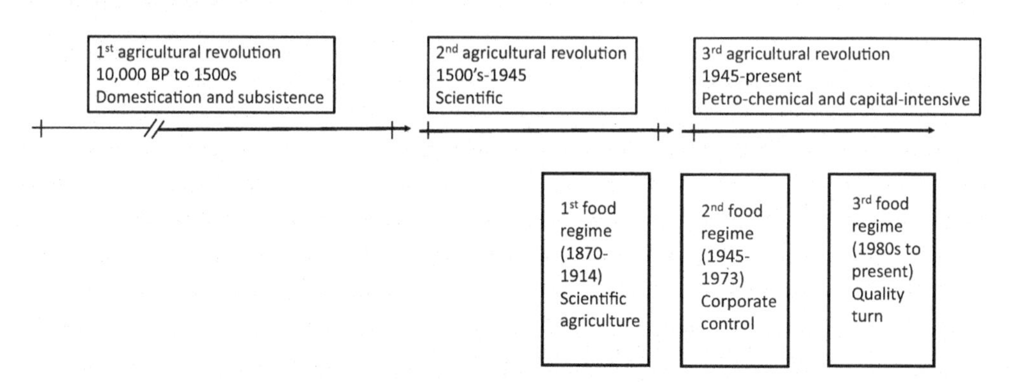

FIGURE 3.5 Agricultural revolutions and food regimes (adapted from Holt-Giménez & Shattuck, 2011)

which European scientists experimented with plant and animal breeding. (Figure 3.5 demonstrates the overlap between the theories of revolutions and regimes.) These innovations accompanied new socio-economic–environmental arrangements, such as the enclosure of common grazing land, which forced peasants off land and into waged work and the application of scientific methods to animal and plant health. Much of the wealth that made such experiments possible was due to the production and export of food from European colonies to markets around the world, much of it produced through slave or coerced labor.

In the middle of the first food regime, the **Industrial Revolution** brought new sources of energy and power to bear on agriculture, such as a steam-powered threshing machine that enabled the **mechanization** of grain harvesting. What was once done by hand in time-consuming and labor-intensive ways (and is still done this way in many parts of the world) could be completed in an afternoon with a few people. Farming families would often invest in and share the equipment given that farm sizes were small enough that no single farmer could afford, or needed, such a machine. This innovation was made possible by the application of fossil fuels, such as coal or kerosene, to the agricultural system, paving the way for widespread mechanization everywhere in the industry. This had the effect of increasing farm size and reducing the number of farms. Because higher yields were now possible, fewer people were needed to run a farm, and the machinery was expensive, thus requiring a large harvest to justify the capital. With mechanization also came **rationalization**, which is a process that increases efficiency through standardization. With mechanized threshing, all the wheat had to be the same height and mature at the same time; therefore, the landraces that were developed over centuries with variability in traits were not suited to the new technologies.

The Third Agricultural Revolution

The so-called **Green Revolution** was an international effort during the **second food regime**, also known as the corporate food regime or the **third agricultural revolution**, spearheaded by the US government and partner corporations to improve productivity of cereal crops, starting with corn, wheat and rice. This effort began around 1960 as an effort to breed and distribute enhanced seeds to countries that were non-aligned in the Cold War. The US feared that growing populations in India, the Philippines, Mexico and Pakistan would lead to a Malthusian famine in these countries. That in turn may make them sympathetic to the USSR who offered food and security. American scientists and corporations thus dedicated efforts to integrate these countries into capitalist markets through agricultural innovations, with an eye toward increasing food exports from those countries to the United States. They focused on producing high-yield hybrid varieties that replaced the traditional seeds or landraces that had been used previously. The perception was that these traditional varieties were low yielding. They were bred to be variable in their height and ripening times and were thus not suited to mechanization. The new hybrids were more uniform, but they required massive inputs of chemical fertilizers and pesticides, as well as far greater water resources, which resulted in the privatization of water into tube wells.

What the scientists overlooked was how the landraces were adapted to many environmental challenges, including drought and pests, and were also used as animal fodder, so the variability in maturation and size allowed them to feed the stalks to animals, such as oxen who provided traction power. The Green Revolution on its public face set out to eliminate hunger, increase agricultural yields and increase technological access in what were seen as underdeveloped countries but resulted in conflict over land, polluted environments and new inequalities (Shiva, 1991). The mechanization of agriculture in many of these countries resulted in **rural outmigration** to cities, where the demand for food increased because those workers no longer grew their own food. The landless, forced off their farms by debt or competition for resources, also could not grow their own food. Meanwhile, the extra calories made available by the Green Revolution increased fertility and lowered infant and maternal mortality, resulting in a large increase in population in the 20th century. Thus, hunger still haunts many of the countries targeted for the Green Revolution, and the countries that benefited the most from the innovations of the second food regime were corporations and the investors who profited from the export of their seeds and their chemicals.

The peak of the second food regime, which emphasized the proliferation of industrial technologies and the cultivation of global markets, includes what scholars like to call the **Gene Revolution**. This innovation in agriculture applies **biotechnology** to plant breeding to make new organisms with unique traits specific to contemporary agricultural problems. Genetically modified (GM) organisms in agriculture have undergone the insertion or deletion of genes, usually across species, that results in a DNA sequence not normally found in nature. The most common species to be used are viruses and bacteria that are bred to exhibit certain traits, which are then transferred to the original organism using recombinant DNA technologies. The three largest GM crops in the contemporary food system by acreage cultivated are corn, followed by soybeans and cotton. Corporate control over the technology intensified with the patenting of seeds. The vast majority of these crops are used in animal feeds or for industrial products. Sugar beets and canola are two big food crops that humans eat in the form of sugar and cooking oils. Genetic modification is also used to make food additives such as xanthan gum and flavorings.

FOCUS SECTION: GM MAIZE IN KENYA

Agricultural biotechnology remains a controversial method of technology transfer from wealthy to poor countries. Analyses of development focused on food insecurity and famine have extensively debated the effectiveness of agricultural biotechnology to create food security in developing countries, and experts are divided on whether or not biotechnologies are beneficial. In Kenya, the growth of agricultural biotechnology has been donor funded and donor led, including donors such as the Rockefeller Foundation and the World Bank, rather than state led for the benefit of the population, in a neoimperial approach to expanding food production globally. Beginning in 1991, Monsanto and USAID sponsored biotechnology in Kenya toward the production of virus-resistant sweet potatoes. Despite initial failures, this project continued and other crop varieties were tested, such as maize, cassava, and cotton. International agencies accelerated research, development, and implementation of biotechnologies due to Kenya's developed scientific and research infrastructure and *lack* of government oversight. Many organizations and corporations produce biotechnologies in the absence of regulation, which leads to a lack of transparency and accountability to farmers and the public who may not want to use GM crops. Figure 3.6 shows Kenyan biotechnologists working on genetic modification on food crops.

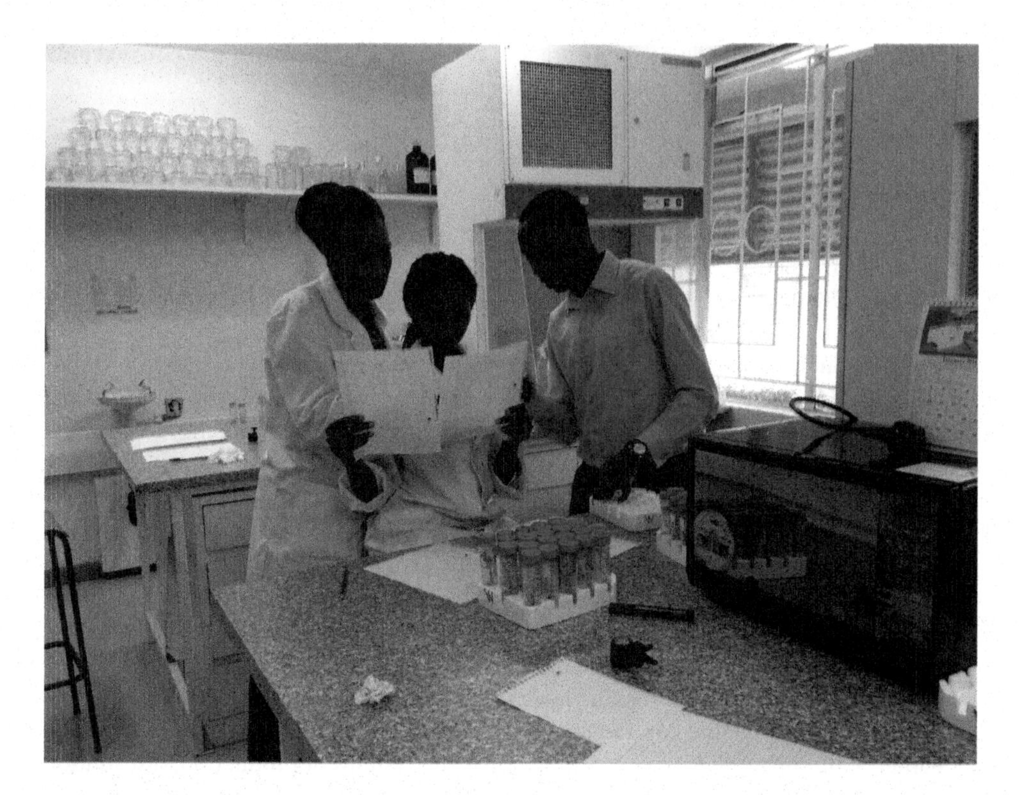

FIGURE 3.6 Kenyan biotechnologists

A Fourth Revolution?

In what scholars like to call the quality turn (Goodman, 2003), or what could be called the **sustainability revolution**, consumers in the late 20th century, becoming aware of the risks to human health, exploitation of labor and threats to the environment from **conventional agriculture**, demanded safer and more ethical food that they could know more about. The beginning of the **third food regime** in the 1980s saw an increase for locally produced, organic products and for food that came from overseas to be grown with more ethical labor practices. The alternative food movements that were born in the era of activism brought new light to bear on production practices, and consumer activism generated new standards for organic and fair trade products. Community-supported and local agriculture also grew as farmers markets returned to cities, offering consumers food from more legible local supply chains. The promise of these movements has faded as organic labeling has become a marketing practice for large corporations, and the production standards are weakened. The cost of producing organic and fair trade (and to some degree local) food means that a **two-class food system** (Hinrichs, 2000) has developed, in which one group is wealthy enough to access purportedly healthier and more ethical food, while another group is not.

Figure 3.7. shows organic vegetables (peppers, eggplants, beans, okra and gourds) for sale in an organic farmers market in Argentina. Organic food production started late and grew slowly in Argentina, until the early 2000s. Argentina produces organic food for export, but there is a small sector marketing produce to higher-income consumers who are relatively young with families and concerned about their health (FAS, 2000). Even in places with a relatively small organic sector, the two-class food system operates to exclude those from lower income classes and their children from safer foods.

SUMMARY

This chapter provided an overview of the agricultural revolutions in which human societies domesticated plants and animals in different regions of the world. The development of agriculture led to new forms of social relationships and new forms of power and control over food production. This in turn led to new innovations and reorganization of societies and the economy. The two major technological changes in agriculture in the 20th century, the use of chemicals to produce food and the manipulation of plant genomes to engineer resistance to chemicals, transformed agriculture in unprecedented ways, in particular, giving more power to corporations and the governments who protected corporate patents. These changes, however, led to consumer backlash, in which organic and fair trade labeling systems and sustainability turns emerged in the later 20th and early 21st centuries. Governance of food production thus shifted from states to third-party, international certifying bodies, which are typically non-governmental organizations. Paying for what is *not* in the food, however, led to higher values and prices in the marketplace. The two-class food system in which the wealthy can consume healthier food than the poor, many of whom grow the food, is one of the central contradictions of the food system, which is discussed further in Chapter 5. But first, the next chapter examines the systems of production, distribution and consumption that characterize the contemporary food system.

FIGURE 3.7 Organic peppers and eggplants in a local market in Argentina (https://upload.wikimedia.org/wikipedia/commons/3/32/Pepperseggplants.jpg)

SUPPLEMENTAL MATERIALS

Key Terms and Concepts: biogeographically inferred, biotechnology, conventional agriculture, domestication, domestication syndrome, first agricultural revolution, food regime, Gene Revolution, genetically modified, green manures, Green Revolution, hunting and gathering, hybrid seeds, Industrial Revolution, landraces, mechanization, open-pollinated seeds, pre-domestication cultivation, rationalization, rural outmigration, scientific agriculture, second agricultural revolution, subsistence agriculture, sustainability revolution, third agricultural revolution, traditional cultivars, two-class food system

Explore More: The Consultative Group for International Agricultural Research (CGIAR) is an international organization dedicated to fighting poverty and hunger with the aim of a food-secure future. The International Center for Tropical Agriculture (CIAT), a research group of CGIAR, developed an interactive map of food origins. Explore the history of a favorite food of yours here: https://blog.ciat.cgiar.org/origin-of-crops/

Recommended Viewing: *Jimmy's GM Food Fight* (2008), *The Man Who Tried to Feed the World* (2020)

Discussion Prompt: The Green Revolution was one of the most controversial and impactful changes to agriculture, changing the way we eat and grow food. From among the many pros and cons, identify (and explain why) which one you think is the biggest benefit and which one is the biggest problem associated with the Green Revolution for our food system. Aim for 100–250 words. Be sure to correctly use at least *one concept* from this chapter in your discussion post and to write it in **bold** font.

Recipe: Cornbread was a staple of the Oneida people in what is now New York State in the US. This recipe captures the traditional method of making it. www.oneidaindiannation.com/the-art-of-cornbread/

REFERENCES

FAS. (2000). *Argentina – Organic food report.* GAIN Report, USDA Foreign Agricultural Service, Buenos Aires.

Friedmann, H. (1987). International regimes of food and agriculture since 1870. In T. Shanin (Ed.), *Peasants and peasant societies* (pp. 258–276). Basil Blackwell.

Goodman, D. (2003). The quality "turn" and alternative food practices: Reflections and agenda. *Journal of Rural Studies, 1*(19), 1–7.

Haas, R., Watson, J., Buonasera, T., Southon, J., Chen, J. C., Noe, S., . . . Parker, G. (2020). Female hunters of the early Americas. *Science Advances*, 6(45), eabd0310.

Hinrichs, C. C. (2000). Embeddedness and local food systems: Notes on two types of direct agricultural market. *Journal of Rural Studies, 16*(3), 295–303.

Holt Giménez, E., & Shattuck, A. (2011). Food crises, food regimes and food movements: Rumblings of reform or tides of transformation? *The Journal of Peasant Studies, 38*(1), 109–144.

Kimmerer, R. W. (2013). *Braiding sweetgrass: Indigenous wisdom, scientific knowledge and the teachings of plants*. Milkweed Editions.

Larson, G., Piperno, D. R., Allaby, R. G., Purugganan, M. D., Andersson, L., Arroyo-Kalin, M., . . . Fuller, D. Q. (2014). Current perspectives and the future of domestication studies. *Proceedings of the National Academy of Sciences, 111*(17), 6139–6146.

Mann, C. C. (2005). *1491: New revelations of the Americas before Columbus*. Alfred A. Knopf Incorporated.

Pascoe, B. (2014). *Dark Emu black seeds: Agriculture or accident?* Magabala Books.

Shiva, V. (1991). *The violence of the green revolution: Third world agriculture, ecology and politics*. Zed Books.

Twyman, J., Useche, P., & Deere, C. D. (2015). Gendered perceptions of land ownership and agricultural decision-making in Ecuador: Who are the farm managers? *Land Economics, 91*(3), 479–500.

Wrangham, R. (2009). *Catching fire: How cooking made us human*. Basic Books.

CHAPTER 4

The Nexus of Production and Consumption

INTRODUCTION

On a trip to New Zealand from Australia a few years ago, the first question I was asked when I arrived in the airport was whether or not I was bringing honey into the country. I've been asked about money, plant material and guns, which are controlled imports for most countries and subject to state power, but I've never been asked about honey. Not carrying honey, I was cleared to enter. I walked through the terminal and saw tiny containers labeled Manuka honey for sale at exorbitant prices. Manuka honey, made from the pollen and nectar of the Manuka tree, native to New Zealand, shown in Figure 4.1, is an extremely valuable export, priced at US$25 for a few hundred grams. The value of the product is in its alleged anti-microbial properties. All honey has such properties, but Manuka pollen is higher in anti-bacterial properties than most plants. And therein lies the reason honey has become a controlled substance in New Zealand.

A major threat to beehives is the bacteria *Paenibacillus* which causes a disease that can kill all the bees in a hive. Foreign honey brought into the country (which bees may eat) could spread the bacteria to New Zealand hives. In other countries, this infection is treated with antibiotics, but Manuka honey would lose its value if antibiotics were used in hives in New Zealand. The government has, therefore, a strictly enforced moratorium on all imports of honey to prevent the spread of infection. This example underscores how **value** is a product of both natural (bees, bacteria and plants) and human systems (markets, cultural preferences, knowledge systems) within a complex system of production, distribution and consumption that is governed by (and often contested by) national and international policy (Nestle, 2013).

This complex set of interconnections in places and regions at multiple scales is a called a **food system**. It is composed of resources, **production**, processing, **distribution**, exchange, **consumption** and waste. It is the process by which natural resources, such as water, soil and seeds, are transformed into edible food products that are consumed by eaters. A food system captures, for example, how wheat is turned into bread. It also, when conceptualized as a **supply chain**, provides information about how something such as a grain of wheat can be turned into a product with exchange value, or a **commodity**. In pre-modern societies, goods were often traded or exchanged as gifts (and these practices still persist in modern economies); it was not until the invention of currency that edible food items could be exchanged through indirect means and also gain value through exchange in a marketplace.

DOI: 10.4324/9781003159438-6

FIGURE 4.1 Manuka flowers (https://commons.wikimedia.org/wiki/File:Common_Tea-tree_(6692434207).jpg)

This chapter examines the nexus of production and consumption as it is shaped by supply chains and the value that accumulates through them.

KEY THEMES AND IDEAS: SUPPLY CHAINS AND COMMODITIES

Supply chains are composed of a series of processes by which raw materials become products that can be bought and sold. Most supply chains have internal supply chains within which other products are produced and integrated into the final product. For example, a loaf of bread will have supply chains for the yeast and salt that are required for the final product and must (usually) be sourced from places other than where the wheat flour is produced. **Value**, or the monetary worth of the product, accumulates in each stage of processing and exchange along a supply chain. The value of the wheat flour (cents per pound) is transformed into the value of a loaf of bread (dollars/pounds per pound) through the application of energy, labor and other inputs. Therefore, the longer the supply chain, the *higher the cost of the final product* must be relative to the *raw materials and labor*. A loaf of bread sitting on the supermarket shelf that has multiple additional ingredients for flavor, preservation and nutritional value is an accumulation of all the value that accumulated through **efficiencies of scale** and **mechanization** to reduce the costs of production and thus make the bread affordable for the consuming public. More about these processes is explained in upcoming chapters, but the inverse relationship between the value of the final product and the costs of the inputs, including labor, is important and will be returned to in this chapter and many others.

The context in which the exchange happens matters. For example, a loaf of bread baked with flour, yeast and water by your neighbor, friend or family member and given as a gift will have a very different value from the loaf of bread in the supermarket. It will have emotional or affective value in addition to likely being too expensive to sell in the current market in terms of the raw materials of inputs purchased at retail prices and the cost of the kitchen space and the value of the baker's time. As a gift, it was always intended to be free or reciprocal, and therefore, the cost of things does not matter much to the exchange. Commodities, however, begin life as objects that are intended to be exchanged. Thus, a food or any other thing becomes a commodity when it passes through a particular, usually monetary, exchange. Appadurai (1988) asserts that commodification is just "one phase in the life of some things" (17) and defines a commodity as a thing in a "situation in which its exchangeability (past, present or future) for some other thing is its socially relevant feature" (13). The bread that was a gift started out life as flour, something purchased as a commodity in the store, but ended up being a gift instead of being sold to a market, which distinguishes it from a commodity only by the context of its exchange. Gifts are also governed differently (usually not at all) from things that people intend to sell, which is discussed later in this chapter. Identifying products as characterized by their "situation" and their social relevance signals that commodities are socially and culturally constructed and, as such, have multiple lives. It also shifts the focus from the product itself (the loaf of bread) to the social context through which a thing travels, which we can call a food supply chain.

Recall the simplified supply chain in Chapter 1, Figure 1.4. Supply chains are composed of the place where the raw materials are created, *the farm*, the place where those raw materials are transformed through manufacturing in *a factory*, where they may also be packaged and then transported through various *distribution* channels to a *market* and then to the consumer's place of *consumption*. Each step in this process will connect people and materials in different places and regions. In a short supply chain, such as a farmers market, this may involve a few miles and a couple of people. In a long or extended supply chain, each step may have an additional internal supply chain, such as for the processing of an additive, and will connect more people, places and regions at different scales. Thus, we can characterize supply chains based on their length, or the number of linkages within them. In a general sense, a long or **extended supply chain** will have a great deal of distance associated with it (i.e., food miles) and associated energy resources contributing to its carbon footprint as well as the social distance between producers and consumers that obscure the conditions of production and manufacture. **Short or local supply chains** promise (but do not always deliver) shorter physical and social distances between consumers and producers and lower carbon footprints, and they almost always involve far fewer inputs and materials from other supply chains.

Commodities gain value through each exchange in the supply chain. For tropical commodities or for highly processed foods, a long or extended food supply chain is required. This implies a great deal of distance between production and consumption or multiple steps in processing, or both. For a granola bar, multiple ingredients, including those with multiple stages of processing, such as flavors or additives, will require extended or long supply chains that are likely embedded in other supply chains. This also brings more value to the product, as capital exchange occurs each time a product changes hands. In **vertically**

integrated systems, one corporation may control the growing of the raw materials, processing and distribution of a food product, thereby the entire extended supply chain. This facilitates two dynamics in the food system: (a) costs are reduced because the product is bought and sold fewer times, but it also (b) concentrates power and therefore choice in the hands of a few decision-makers. In most supply chains, there are many producers and many consumers but very few processors or distributors. Where there are few decision-makers, there is tremendous power. For example, if there are only a few corporations that have invested in costly shipping infrastructure for tropical commodities (i.e., planes or ships), those companies will control the logistics *and* the price of the shipping. Not all companies will have the kind of financial power to purchase those sectors of the supply chain for themselves, so they will be a "price taker" on transportation; therefore, those who control that part of the supply chain will be the "price makers" for the whole system. The many producers and consumers will have very little say in the price. They can only control what they are willing and able to pay (consumers) or are willing to be paid (producers) for their product. Shorter or local supply chains offer opportunities for consumers to partner with producers and have more influence on production practices.

Short and Local Food Supply Chains

Short or local supply chains, while similar, have different meanings in practice (Kneafsey et al., 2013). A **local food system** is production, processing, exchange and consumption within a defined geographic area. "Local" may mean within the scale of a city or a county or a state and is generally accepted to be between 20 and 100 km in radius. It is not well defined but easily translates for both producers and consumers into "local equals good." **Short food supply chains** reduce the number of links in the chain and focus on direct marketing, such as with producer-only farmers markets. While logistically this may mean a local food system, they can also include direct sales at a distance, such as with wild salmon fishers who transport catch back to their home markets or direct marketed coffee that is purchased from farmers by a roaster and sold at the roasting facility. Taken together, these two kinds of supply chains can render the food system more legible and more accountable to consumers. While they may provide multiple benefits, a study of short food supply chains in Paris found that, like many urban agriculture initiatives, land was in short supply and threatened with development (Aubry & Kebir, 2013). Figure 4.2 shows a typical local food market in Dublin, Ireland. The seasonal nature of produce is apparent in the abundance of greens, tomatoes and other summer crops. In this case, the woman vending was not the farmer but someone hired by an organization of farmers to market their collectively produced food.

There are three main types of local and short food supply chains. The first is community-supported agriculture, or a subscription scheme as described earlier. This involves consumers receiving a box of produce or other products, including cheese, honey or meat in addition to fruits and vegetables in exchange for a monthly or semi-yearly payment. There are no intermediaries in this supply chain, particularly if consumers receive the box on the farm. The second is an on-farm direct marketing situation where consumers buy individual products such as milk, eggs, meat or fruits and vegetables at a farm stand. In some cases, farm stands operate on an honor system, and no interpersonal interaction

FIGURE 4.2 Farmers market in Ireland

happens at all. The third is an off-farm direct marketing situation, such as a farmers market, where farmers bring produce to a central location, and consumers meet them there. Some farmers markets are "producer-only," meaning that no one but the farmer may sell the products. Some studies have shown that farmers who are successful at this approach often diversify their markets into multiple such markets, including food hubs and food marketing cooperatives, thus making a more viable farm income.

Each of these supply chains intentionally keeps the number of exchanges at a minimum, which has a number of documented benefits. The first and primary benefit is that farmers keep most of the consumer's dollar that is spent on food. The other benefits tend to be less tangible, but some studies indicate benefits for rural development, environmental health and social cohesion (Renting et al., 2003). Critics of short and local food supply chains identify them as exclusive, whitened spaces that benefit the already wealthy and privileged the most (Guthman, 2008). Alternatives that are less exclusionary often tend to be anti-capitalist as well, such as community garden projects where food is grown in unused urban lots and distributed to the local neighborhood. While local and short supply chains may redress some of the problems of food insecurity and farmer incomes, one place where scholars disagree on the benefits of local and short food supply chains is on the environmental impact of long supply chains. The conventional wisdom has been that the higher the food miles, or the cumulative physical distance in the supply chain between production and consumption, the worse is the environmental impact, particularly when it relates to the release of greenhouse gases that drive the earth's rising temperatures, fueling climate change. This logic breaks down when examined from the lens of efficiency. According to some studies, the carbon footprint of a farmer taking a truckload of green beans to a farmers market is larger per unit of green beans than a much larger volume of French beans grown in Burkina Faso and airlifted to Europe. While the food miles may be longer, the volume of production makes it more efficient. This comparison may work for out-of-season temperate produce, but it cannot be turned into a useful comparison for something like an entire meal composed of multiple ingredients sourced from multiple regions and/or countries.

Geographers and others use the concept of the **foodshed** to expand the idea of food miles and also to capture more accurately and in more detail from where food comes. A foodshed, according to Feagan (2007), is a network of the flow of food produced for a particular population. Like a food supply chain, it captures all aspects of production, distribution and consumption but displays it geographically. It is based on the idea of a watershed, which is the flow of water in a particular region, but is "socio-geographic" in nature and captures human and natural systems. The size of a foodshed can vary widely, and in the case of the global diets of most consumers in the developed world, the foodshed includes the entire world. Figure 4.3 shows a McDonald's burger and fries and captures

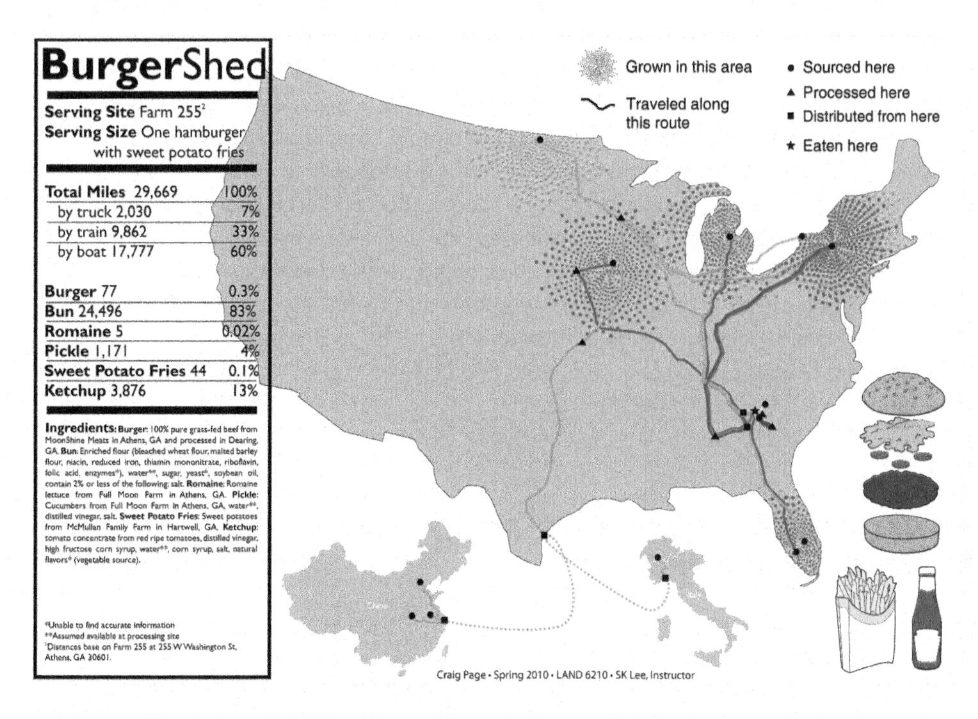

FIGURE 4.3 The McDonald's BurgerShed (Craig Page, used with permission)

a few key ideas about food systems. The first is the wide geographic extent of most food ingredients that accumulate much more miles than simply the distance between production and consumption. The second idea is that unless there is an audit system in place, most consumers cannot know the exact location of food production, and it can only be reduced to a likely source, such as a potato- or wheat-producing region.

FOCUS SECTION: FOOD WASTE

Most discussions of food supply chains do not include food waste. **Food waste** is an enormous problem, primarily in the world's wealthiest countries, and an estimated one-third of food is lost or wasted. This is discussed in further detail in Chapter 5 as a central contradiction in the food system, but there are several innovative approaches to alleviating food waste, which include intercepting it and reclaiming it. Part of the reason for this is strict food safety standards that limit how foodstuffs can be distributed after their packaging is opened. Campus Kitchens is a nationwide program in the US that focuses on reclaiming usable food that would otherwise be thrown out from college dorm cafeterias. For example, if a bag of 12 hamburger buns is opened and only one is used in a given time frame, the other 11 will be thrown out. Students prepare meals using that donated food, as well as food from local food banks, restaurants, grocery stores and farmers markets. Then, students deliver meals free of charge to individuals and agencies in the school's neighboring community, such as homeless shelters, food banks, soup kitchens and individuals or families in need of food assistance. Another problem with food waste is spoiled food that could be composted but is sent to landfills instead, contributing in various ways to the carbon footprint of the food system. Composting is practiced in places where people have access to the space for it, but many people, especially in urban areas, lack the capacity to compost food waste. Community composting programs pick up food scraps at the curb and take them to composting facilities, where they are turned into soil for community gardens. Larger-scale operations can include event composting, waste-diversion and composting solutions for businesses ranging from restaurants and coffee shops to hotels, stadiums and manufacturing. To lower the carbon footprint of the businesses, some community composting programs use electric vehicles and bicycles to pick up and transport food waste.

CONCEPTUAL ENGAGEMENT: FOOD SUPPLY CHAIN GOVERNANCE

Consumers are often given the illusion of power and choice in the food system by being confronted with a dizzying array of options in retail markets. Granola bars, for example, through their branding, advertise distinctive traits: high protein, low fat, high fiber, low sugar and so on. But if the labels were read for manufacturer name, the hundreds of apparent options in the store would be reduced to about five. The vast majority of the food produced in the global food system is controlled directly or indirectly by the five largest (by revenues) food corporations: Cargill, ADM, Nestle, Sysco and JBS (a Brazilian company producing factory-processed meats). In a global food system, each product manufactured in an extended supply chain must pass stringent food safety standards that require global **governance**. Each country or region, such as the European Union, may have its own laws, but these must be harmonized with food safety laws in the place of consumption, which requires coordination between actors. Recall from Chapter 2 that three different types of actors work to govern food systems. In the global food system, this is largely a coordinated effort between states and corporations, and consumers and producers have very little influence, in spite of their greater, relative numbers.

EXAMPLES

In what follows, two global supply chains and one local food system are discussed with a specific emphasis on how they are governed.

Assam Black Tea

Global markets for tea have long been associated with British colonialism in India. Along with sugar as a carbohydrate made cheap by slave labor on the Caribbean sugar plantations, caffeinated black tea was a stimulant used by industrial workers in European factories. Tea has become a symbol of Britishness and the good life, which belies its history of forced labor and colonialism at its origins. The major tea growing areas of India include states in the northeastern part of the country, some of which is contested territory and marked by civil conflict. This region includes the state of Assam, where the majority of the world's black tea is grown. India is the second largest tea producer behind China and exports several million tons of black tea yearly to the large tea-consuming countries of the UK, Australia and some European countries. Labor issues on tea plantations have long been a significant problem but largely hidden from consumer view due to the length of the supply chain. They include forced labor, workplace violence, child labor and wage theft. Major tea brands, such as Lipton, which are owned by the giant food and beverage corporation Unilever, in recognition of this fraught history, went so far as to stamp their products "slavery free." According to reports in 2018 and 2019 however, these workplace conditions persist. Governance of working conditions is generally not a priority of global governing bodies, such as the World Trade Organization or the Food and Agriculture Organization of the United Nations, and labor regulation is often tackled on the scale of the individual nation-state. Consumer labeling efforts can do little without oversight at the global scale, and when capital accumulation is the driving force in the supply chain, employers will often suppress wages to increase profits. The most vulnerable, women and children, who make up the majority of tea pickers suffer the most. Figure 4.4 shows a woman picking tea in the Ceylon tea plantations of Sri Lanka.

FIGURE 4.4 Woman picking Ceylon tea in Sri Lanka (https://commons.wikimedia.org/wiki/File:Sri_Lanka,_Tea_plantations,_Nuwara_Eliya,_Picking_tea_leaves.jpg)

Cuban Honey

Following the collapse of the Soviet Union, Cuba's Soviet-subsidized industrial agriculture collapsed as well, which led to deep austerity during what was called the "Special Period in Peacetime." In order to build a more self-sufficient agricultural sector moving forward, the Cuban government adopted agroecological principles as the basis for food. Today, Cuba is globally recognized as an agroecological leader for its encouragement of sustainable food production. It has provided policy and material support for agroecology through state-based mandates. These include restrictions on the use of agricultural chemicals, land reform and redistribution of fallow land to small-scale farmers and cooperatives, and significant investment in agroecological research and extension programs. Honey is a difficult product to govern because bees forage widely and may encounter chemicals while doing so. In Cuba, however, chemicals are tightly controlled, making organic certification possible. Organic honey is growing as a high-value Cuban export, ranking 12th in value among Cuban exports in 2016, at US$14.3 million. In recent years, Cuba exported over 97% of its organic honey production to eight European countries, including a large market in Switzerland. Organic standards for imports to Europe are established and certified by a German third-party agency, KIWA BCS Öko-Garantie. The state-based reforms in Cuba enable a chemical-free product to be exported and facilitate the growth of a niche product sector. This is only made possible, however, by a non-profit mode of governance in the form of a third-party organic certifier located in another country.

Local Food Systems: Australia

In the late 20th and early 21st centuries, frustration with the social and environmental problems in the global food system led many consumers to demand certified organic and fair trade products and to consume food at a more local scale, where the ability to know about the conditions of production were greater and the accountability for the "bads" of the production process could be dealt with in a local governing context. This resurgence of local food systems came in the form of local farmers markets and **community-supported agriculture** schemes, a system where a consumer buys a share in a farm and gets a box of food weekly. These were largely driven in an ad hoc way by consumer demand and without a lot of government involvement or oversight, even in terms of food safety. As local agriculture began to show promise for an environmentally sensitive rural economic development strategy providing multiple ecosystem benefits, governments took notice and became involved, to varying effects.

An initiative in Victoria, Australia, details how local agriculture can provide ecosystem services, such as flood control, recreational opportunities as well as guaranteed access to fresh food in urban areas. Peri-urban agriculture, or the cultivation of food adjacent to or within metropolitan areas, can include "green wedges" that promote biodiversity while producing food with small, medium and large integrated crop-livestock systems that can also be used as agritourism. Nearby waterways can provide opportunities for aquaculture, while providing flood control and recreation. They can also reduce the urban heat island effect, provide corridors for pollinators and provide aesthetic benefits for urban residents. Agroforestry initiatives can provide food while sequestering carbon and resources for sustainably harvested timber. While understanding that cities will be dependent on other supply chains for all their food needs, the initiative reveals how producing food in or near the city has multiple benefits in terms of ecosystem services and recreational opportunities for residents (Dubbeling et al., 2017).

FOCUS SECTION: FOOD AND CULTURE

Culture is its own topic that could take an entire semester, but it is worth discussing here briefly. The value of things (seeds, land, labor) and an overarching political system of governance largely determine what is produced and where, and therefore the supply of food. The demand for certain foods is largely driven by preferences based on taste and cultural appropriateness, which are also shaped by power in the food system. Geographer Don Mitchell (1995) argues that culture is just politics by another name, and nowhere is that more evident than in our food system. What we eat and why is a product of empire, colonization and capitalism. Even food traditions that survive enslavement and forced migrations, such as rice and benne (sesame) that women braided into their hair, have survived because there was some economic or political value in their preservation. Another example is the concoction known as lutefisk, that my ancestors brought to America with them from Norway. Lutefisk is cod soaked in lye, dried and then reconstituted by boiling and served, if you're lucky, on Christmas Eve with lots of bacon to mask the taste and the smell. Legend has it that Viking fishermen hung their cod to dry on tall birch racks. When some neighboring Vikings attacked, they burned the racks of fish, but a rainstorm blew in from the North Sea, dousing the fire. The remaining fish soaked in a puddle of rainwater and birch ash for months before some hungry Vikings discovered the cod, reconstituted it and had a feast. Norwegians brought this fish with them on a long migration to settle in the US as colonists, which displaced and killed millions of Indigenous people. This tradition has been retained in the place that is now Minnesota, partly due to a shared symbol of European history and supremacy, and clearly not because it tastes so good. The traditional shared harvesting of wild rice that used to take place in Minnesota is not so much a part of the settler culture. So, while culture may influence what we like and why we like it, in many ways power in its coercive or symbolic forms has far more influence on what we eat.

SUMMARY

This chapter provides a general overview of food systems and their associated vocabularies that are in use throughout the book. This chapter underscores an integrated approach to studying the creation and governance of food supply chains from field to fork and beyond through the lens of geography. Food systems serve as the nexus of production and consumption where the natural resources, labor and energy inputs of the food system generate consumable products. Food moves along global and local supply chains and is governed with varying degrees of power by states, corporations and civil society. Producers and consumers, while great in number, have very little power in global supply chains, often resulting in human rights abuses, such as those seen in the Assam tea production, by the actors who hold the most power in the supply chain. Consumers through demanding quality or accountability, such as those consuming Cuban honey, can exert power through non-profits and civil society working with the state. Short food supply chains, often embedded in foodsheds, offer some benefits to workers and producers and are far more transparent and legible to consumers. Finally, waste, as an often underdiscussed product of food supply chains, is both an environmental and a social problem, which is further discussed in the next chapter.

SUPPLEMENTAL MATERIALS

Key Terms and Concepts: commodity, community-supported agriculture, consumption, distribution, efficiencies of scale, extended supply chain, food system, food waste, foodshed, governance, local food system, local supply chain, mechanization, production, short food supply chain, short supply chain, supply chain, value, vertically integrated

Explore More: Unilever claims its tea is slavery free, but some reports suggest otherwise. A global campaign to certify foods and other products as slavery free can be examined in more detail here: https://beslaveryfree.com/

Recommended Viewing: *Dirt! The Movie* (2009), *Fresh* (2009), *Before the Plate* (2018)

Discussion Prompt: Food writer Michael Pollan claims that the way we eat has changed more in the last 50 years than in the last 10,000 years. In what ways have food systems changed? How have they stayed the same? Give one example of each from your own or a global diet. Aim for 100–250 words. Be sure to correctly use at least *one concept* from this chapter in your discussion post and to write it in **bold** font.

Recipe: *Chai* is the Hindi word for "tea," and it is easy to make at home from spices that might already be on hand. It will make the kitchen smell amazing. Serve with milk and honey. www.epicurious.com/recipes/food/views/homemade-chai-201226

REFERENCES

Appadurai, A. (Ed.). (1988). *The social life of things: Commodities in cultural perspective*. Cambridge University Press.

Aubry, C., & Kebir, L. (2013). Shortening food supply chains: A means for maintaining agriculture close to urban areas? The case of the French metropolitan area of Paris. *Food Policy, 41*, 85–93.

Dubbeling, M., Bucatariu, C., Santini, G., Vogt, C., & Eisenbeiß, K. (2017). *City region food systems and food waste management: Linking urban and rural areas for sustainable and resilient development*. Deutsche Gesellschaft für Internationale Zusammenarbeit (GIZ) GmbH.

Feagan, R. (2007). The place of food: Mapping out the "local" in local food systems. *Progress in Human Geography, 31*(1), 23–42.

Guthman, J. (2008). Bringing good food to others: Investigating the subjects of alternative food practice. *Cultural Geographies, 15*(4), 431–447.

Kneafsey, M., Venn, L., Schmutz, U., Balázs, B., Trenchard, L., Eyden-Wood, T., . . . Blackett, M. (2013). Short food supply chains and local food systems in the EU. A state of play of their socio-economic characteristics. *JRC Scientific and Policy Reports, 123*, 129.

Mitchell, D. (1995). There's no such thing as culture: Towards a reconceptualization of the idea of culture in geography. *Transactions of the Institute of British Geographers*, 102–116.

Nestle, M. (2013). *Food politics: How the food industry influences nutrition and health* (Vol. 3). University of California Press.

Renting, H., Marsden, T. K., & Banks, J. (2003). Understanding alternative food networks: Exploring the role of short food supply chains in rural development. *Environment and Planning A, 35*(3), 393–411.

CHAPTER 5

Food System Contradictions

INTRODUCTION

Lamb, or sheep meat, is a luxury food for many people in the developed world. Sheep are highly adaptable and hardy and can be raised in nearly any environment, but Australia and New Zealand dominate global supply chains for sheep meat. Sheep were brought to Australia by British colonizers, and their impact on the carefully tended soils and crops of the Indigenous people was devastating to their food security. Indigenous people who live where sheep are produced have some of the highest rates of food insecurity in the world at 22%, compared to 4% of non-Indigenous Australians. That this global luxury food is produced where local people starve intersects with the fact that lamb can be easily, cheaply and sustainably produced virtually anywhere in the world. Figure 5.1 shows a flock of sheep living within a few miles of my home in Georgia. Sheep (and their cousins, goats) have less need for water and added calories from grain compared to cattle, and they have a higher-quality meat. So, a geographer would ask, what explains the dominance of one region of production in a supply chain and the socio-economic disparities that shape the place of production?

This chapter elaborates on the relationship between inequities and injustice in the food system that have received scholarly attention in the late 20th and the early 21st centuries. This chapter focuses on the contradictions of the modern food system that produces too much food in one place or region and not enough or not the right kind in another, thus generating surpluses and waste on one hand and hunger and malnutrition on the other. This chapter engages with histories of colonialism and the perpetuation of capitalism as productive of food insecurity through the interconnectedness and interdependence of place, region and economies. As elaborated on in Chapter 4, the problems that commodification of food poses for food security are not the commodities themselves, but the process, arguably a capitalist one, that produces them, and through which they are distributed. The "socially relevant feature" of food as a commodity is its value in *exchange*, rather than its ability to provide sustenance and nutrition through its *use* as food.

In what follows, two central contradictions of the industrial food system are outlined. The first is the situation of chronic hunger in one place, when overproduction occurs in another place, generating food waste. Chronic hunger, therefore, is not the result of shortages or undersupply but a distribution problem where some have too much food available to them and others have too little. This process is spatial and a function of power in the

DOI: 10.4324/9781003159438-7

FIGURE 5.1 Katahdin/Dorper cross sheep in Georgia

food system, which follows paths of capitalist development. The second contradiction is the two-class food system that is generated through efforts to fix the environmental-social inequities in the food system. This contradiction, like the first, is characterized by some having healthy food available to them, while for others, often in the same place, healthy food is neither available nor affordable.

CONTRADICTION 1: UNDERNUTRITION/OVERNUTRITION

Chronic hunger is a situation that develops when people do not receive adequate nutrition and calories to sustain a healthy life over a long period of time. One in nine people in the world suffer from it, and it is the leading cause of child death and stunting in the world, from both maternal undernutrition and food insecurity. **Food insecurity** is a state in which people do not know if they will be able to consume their next meal. This is either because food is not available or because they cannot afford to buy the food that is available to them. In addition, food insecurity includes not being able to control if the food available to them will be adequately nutritious or culturally appropriate. More than 800 million people live with either food insecurity or chronic hunger. The World Food Programme estimates that number will increase to 840 million by 2030. Hunger is most prevalent among the least economically developed countries as well as places experiencing conflict, such as Chad and Afghanistan. Most measures of food insecurity and hunger at the national level obscure the scale of experience however, and while Australians in general do not experience hunger, more than 4 million people, primarily Indigenous women and children, have experienced food insecurity in the last year.

FOCUS SECTION: FOOD SWAMPS

The term **food desert** was used in the 1990s by the US Department of Agriculture to characterize a place, usually an urban neighborhood, at which access to affordable or nutritious food is limited for certain groups, often racial minorities or the poor. Food deserts are measured by the distance between a person's home and the nearest grocery store. Distances vary according to place, but for urban areas between a half to one mile (1–2 km) means food availability is too low. In rural areas, the distance is 10 miles (20 km) (Walker et al., 2010). This distance is also heavily influenced by the racial and ethnic makeup of the neighborhoods, with predominantly White communities having greater access regardless of income. This is largely attributed to the racist decision-making of large-scale retailers who fear loss of market value in neighborhoods composed predominantly by people of color. Researchers examining the landscapes of food deserts, however, discovered that in between home and grocery store are convenience and liquor stores and gas stations offering high-sugar, high-fat and heavily processed foods. They recast the term *food desert* to **food swamp** to better capture the way in which some neighborhoods are saturated with unhealthy food (Rose et al., 2009). Critics of the terms argue that the analogy is not apt, since swamps (and deserts) are generally healthy, life-giving places. They encourage an ecosystem metaphor that considers the multitude of unhealthy decisions that impact the environment, farm workers and other food system workers that created the disproportionate access to unhealthy food in the first place (Elton, 2019).

Meanwhile, food waste is a large and growing problem globally, with the wealthiest countries throwing out the most food. The Food and Agriculture Organization of the United Nations estimates that around one-third of the world's food was lost or wasted every year. **Food loss** is the decrease in the quantity or quality of food resulting from decisions and actions by food suppliers in the chain, up to, but not including, the retail level. This refers to food that is damaged, rotted, imperfect or otherwise not suitable to be marketed. **Food waste** refers to the decrease in the quantity or quality of food resulting from decisions and actions by retailers, food service providers and consumers. Food is wasted because it is beyond its "use by" date, rejected by consumers or unused in household kitchens, delis or restaurants. The US is estimated to waste nearly half its food supply, keeping food out of the hands of people who need it. It is also a waste of productive resources, such as land and water, and energy used to produce and transport discarded food to landfills. Food loss and waste occur for many reasons, with some types of loss – such as spoilage

– occurring at every stage of the production and supply chain. Between the farm gate and retail stages, food loss can arise from problems during drying, milling, transporting, or processing that expose food to damage by insects, rodents, birds, molds and bacteria. In the retail sector, equipment malfunction (such as failed cold storage), over-ordering and culling of blemished produce can result in food waste. Consumers also contribute to food waste when they buy or cook more than they need and choose to throw out the extras (Buzby et al., 2014).

Kate Raworth (2017), in her book *Doughnut Economics*, estimates that it would only take 3% of the global food supply to feed the 13% that are undernourished and that this could be easily met with only 10% of the food that is wasted. The fact that some have so much they throw it out and some have so little they starve, sometimes in the very same place, is a central contradiction in the food system. Raj Patel (2005), in his book *Stuffed and Starved*, argues that the Green Revolution did not eliminate hunger but rather imposed a restructuring of the food system, dominated by global markets, and thus reorganizing who is hungry, where and why. New forms of malnutrition today coexist on a global scale: **undernutrition** and **overnutrition** are now thought to be equally dire threats to human health. Figure 5.2 shows indicators of undernutrition that include anemia (or iron deficiency) in women of reproductive age and stunting of children under age 5 years. In some cases, these exist alongside adult overweight indicators, as in the case of Saudi Arabia and China. This triple burden of **malnutrition** reflects deep inequalities among the country's population where some have too much and others have far too little to eat. Most of sub-Saharan Africa, South Asia and Oceania, which were recently independent

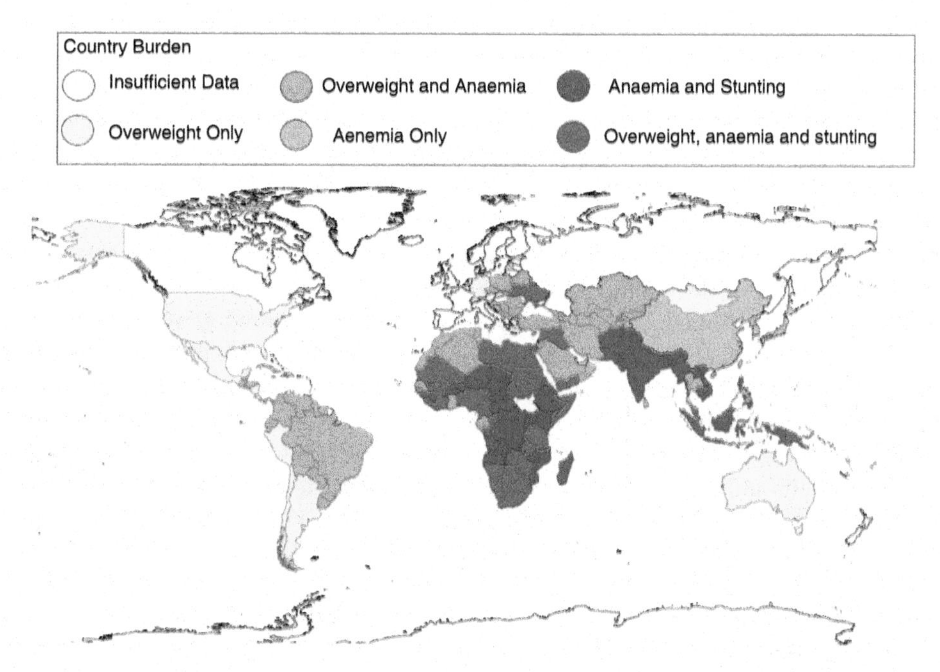

FIGURE 5.2 The multiple burdens of malnutrition (data source: Global Nutrition Report, 2020)

colonies of European powers, have the highest levels of undernutrition. Countries indicating insufficient data either do not collect data on these indicators or do not release them to the public.

For example, Zimbabwe is a country that is characterized by overweight, anemia and stunting in children, presenting the triple health burden related to food in the country. The Global Nutrition Report in 2020 indicates that nearly 30% of women of reproductive age in Zimbabwe do not consume enough iron, and 24% of children are vulnerable to slowed growth related to diet deficiencies. This figure is lower, however, than the African region, and the country reports that it is on target to slow the growth of both anemia and stunting. Zimbabwe, however, is also characterized by a higher than average rate of obesity in women at nearly 28%. The figure is lower than average for men (5.6 versus 9.1 for all of Africa). While obesity is a poorly understood and inaccurately measured metabolic syndrome, high rates of obesity are closely correlated with low-quality diets and high levels of inequality. Zimbabwe's measure of income inequality (score of 50 in the Gini index) suggests some of the highest inequality in the world. This correlates to 1% of the population holding 50% of the wealth of the country (Treanor, 2015). Inequality is the single largest driver of contradictions in the food system, and it follows from the power of capitalist markets to shape who eats what and where.

The Green Revolution, far from preventing hunger and food insecurity, simply reorganized the agricultural production system on a global scale. What had once been a primarily domestic affair, undertaken by individual nation-states and in former colonies to be exported to their former colonizers, the Green Revolution made the production of staple foods, such as grain, an international priority for some nation-states. The Green Revolution undoubtedly increased calories per capita and decreased the general incidence of famine (although starvation is still a reality for many, including people currently living through civil conflict). The increase in food production in the Green Revolution target countries, India, Pakistan, Mexico and the Philippines, allowed those countries to shift from being net importers of food to net exporters. For those who benefited from this shift, such as the large landholders and exporters, there was a rise in income and standard of living. The general, overall increases in national-scale food security that allowed some grains to be stockpiled as commodities contributed to the overall wealth of the nation, although the distribution of those benefits was not enough to lift everyone out of poverty or alleviate hunger, especially for the poor and landless.

The Green Revolution introduced hybrid seeds that required fertilizer and other inputs, such as pesticides, which had to be purchased at the beginning of the growing season, before crops had been sold, and put farmers in a cycle of indebtedness, many for the first time. Remember that in subsistence agricultural systems, seeds were saved from the previous year, animals provided fertilizer and traction power and local varieties were adapted to pests and other threats, such as drought. With the genetic diversity they held, some varieties could withstand disaster, and other varieties could not. The hybrid varieties were all genetically similar, so if one succumbed, all were vulnerable, requiring farmers to adapt the environment to the seed, rather than the other way around. This led to an increase in pesticide use and a cycle of increasing pest resistance and pollution. Hybrid varieties also required heavy watering, and farmers dug tube wells rather than relying on community wells, leading to depleted groundwater and salinization of soils. The shift to mechanization meant that few people were needed to grow the same crop, which led to the migration of rural surplus labor to cities, which led to land concentration as farmers lost or sold their farms.

Recall from Figure 1.2 that as agricultural productivity increases, people migrate to cities, and the increased food supply allows those who can purchase it to have larger families. This led to the growth of megacities in developing countries, which requires an integrated system of food distribution and the ability of people to pay for food. The rural laborers who migrated to cities now need to buy food, and if they do not have a job or money, they go hungry, deepening inequities in those communities. In a cruel twist, food costs the most for the poorest, who spend more of their disposable income on food (up to 75%) than the wealthiest (less than 10%). The concentration of farmland and transition from subsistence to cash crops meant that the poor in the countryside also went hungry. Malthus was worried about population growth outstripping supply and so was Norman Borlaug, the architect of the Green Revolution, who worried that a hungry people were more likely to revolt and overthrow governments. But the population kept growing, and hunger did not stop. Thus, there remains a vexing chicken and egg question about whether agricultural surpluses actually create population growth and therefore hunger when food supply is not available to that new population. What Malthus and Borlaug maybe should have been worrying about is how to control population growth through providing women with the means and options to control their family size. This strategy, now in use in many places, including India, would not have left us with more people, ongoing hunger, deepened inequality and polluted environments.

CONTRADICTION 2: TWO-CLASS FOOD SYSTEM

In the second half of the 20th century, a few decades after the Green Revolution and multiple investigations into the damage to human and environmental health caused by agricultural chemicals, consumers began demanding a safer and more legible food supply. The emergence of organic agriculture in the 1980s and its widespread adoption as a federal program in the 2000s signaled a change on the part of both producers and consumers to reject environmentally damaging practices, although sales of certified organic products remain small in comparison to total food sales. Certified organic production requires 3 years of farming without the use of petroleum-based chemicals, no use of genetically modified seeds and other long lists of "must have" and "must not apply" that are specific to each certifying agency. The development of standards for fair trade similarly signaled a rejection, largely by consumers, of unfair labor practices and unfair prices for tropical commodities, such as coffee and chocolate. Sales of Fair Trade Certified products are higher in Europe than in the US, and such products command a growing market share globally. Figure 5.3 shows a box of certified organic produce from a farm in central Pennsylvania, destined for high-end restaurants in Washington, DC. The labels on the box signal to the consumer that the product was produced with a set of standards ensuring environmentally sound practices, thus justifying the extra expense for the product and contributing to the development of a **two-class food system**.

The ongoing market-based relations inherent to organic and fair trade – that is, consumers still have to buy it and generally pay more for it – ensured that certified organic products would be developed by multinational corporations whose CEOs wanted a share of the market. Many well-known organic producers (such as Cascadian Farm) that started out as independent companies are now brands owned by one of the world's largest corporations (in the case of Cascadian Farm, General Mills). Consumers are willing to pay more for a certified organic or fair trade product in the belief that they are "doing good," but they have little or no control over how the benefits are distributed or accrued on the other end of the long, illegible to most, supply chain. Far from addressing the failures of the market to ensure justice for consumers and producers, organic and fair trade have scaled up the governance of food from the state regulatory apparatus to supranational non-governmental organizations that govern with voluntary auditing systems. These consumer-driven and market-based initiatives and their codification into labels and certifications only made organic and fair trade agriculture "safe for capitalism" (Guthman, 1998, p. 150), facilitating the development of inequalities in terms of who produces and who consumes these products.

Figure 5.3 shows certified organic cucumbers and fava beans grown in Pennsylvania and destined for high-end restaurants in Washington, DC. The farmers who grow the food told me that they could not afford to buy the food they grow in those markets and that they sold it through a local marketing cooperative to high-end restaurants because the rural communities they live in would not support the market that they need in order to make a profit. The production of inequality that persists in capitalism appears even in markets characterized

FIGURE 5.3 Certified organic vegetables

by local supply chains or **embeddedness** in the community, leaving upper- and middle-class consumers with more power and privilege than farmers or lower-income consumers (Hinrichs, 2000). Local food systems also trend toward a "defensive" (Winter, 2003) or "unreflexive" (DuPuis & Goodman, 2005) stance against global capitalism without examining how capitalist orientations in local markets reproduce the inequality that embeddedness set out to disrupt. The primary reason, however, that tomatoes are more expensive at the farmers market than they are at the grocery store is not because local farmers are making so much more money but because the industrial food system is heavily subsidized by federal governments, including subsidies for the fossil-fuel industry, without which the industrial food system would collapse, therefore artificially lowering the price of food.

This has been part of a general trend toward **neoliberalism**. Neoliberalism is a form of governance that facilitates accumulation in the private sector. The belief system behind neoliberalism is that markets can produce public goods better than governments can. It reveals the false binary or separation between markets and states that is often referred to as "free markets" or laissez-faire economics. Nation-states have always been interested in developing their economies, whether they be socialist or capitalist. In some cases, the government takes a stronger hand in regulating and developing industry and markets (i.e., nationalizing energy reserves and selling electricity directly to consumers at cost). In some cases, it takes less of a strong hand through privatization and deregulation (i.e., privatizing energy production and distribution but setting price ceilings). In some cases, the state also creates the conditions for the facilitation of private gain, often at public expense (i.e., removing price ceilings for utilities, meaning the consumer must pay whatever the company charges for electricity or water). The philosophy at play here is that companies will compete with each other for the consumer dollar, thereby lowering the price for the product. In the absence of competition and with the growth of monopolies, this philosophy ceases to apply. In the case of the food system, policy and governance are set up to benefit the industrial and financial sectors of the agricultural system through deregulation and privatization of resources, subsidies that directly encourage the production of crops that benefit multinational companies and the failure to break up monopolies that dominate the food system.

Capital accumulates in the supply chain where the actors are few and decisions about price, cost and logistics are controlled by a few large companies. These few decision-makers determine farm worker and other food system worker wages that are often so low, people cannot afford to buy food (McMillan, 2012). Those low wages allow profit to accumulate further up the supply chain in the form of CEO salaries and shareholder stocks in the multinational companies that dominate the supply chain. The political clout this economic power affords these multinationals also empowers them to lobby federal governments and influence policy on subsidies, patents, cheap food and labor practices. The subsidies that encourage certain kinds of commodity production also generate surpluses of foods that are used as tools of foreign policy and turned into cheap, readily available calories in the form of sugary, high-fat processed foods. In this way, economic power begets more economic power because each of these policies contribute to the accumulation of more profit. They encourage more farmers to grow the patented crops that farmers must pay royalties on, and the processing of foods that are unhealthy but appeal to consumers escalates at ever more diminishing wages.

So, before the Green Revolution, a two-class food system exists with wealthy, often settler elites enjoying sufficient nutrition often at the expense of the poor and landless who were unable to purchase adequate food for themselves and their families. After the Green Revolution, the overproduction of cheap food means more poor people can buy it, but the food is unhealthy and contributes to social and environmental injustice and much is wasted; on the other hand, social movements created expensive, "ethical" food that is ostensibly cleaner and better, but only those with high enough incomes can buy it. What has remained the same over the course of the 20th century and into the 21st century is the market-based rationality of the food system, in which food is for sale and its primary purpose is not to nourish but to facilitate the accumulation of capital in the private sectors of industry and finance. This has been accomplished through the dispossession of Indigenous people, the enslavement of workers and the exploitation of the working class. It should come as no surprise then that globally, the places where food insecurity persists are former colonies and the places where people work in the primary sector to produce food for the world. They are hungry because they lack the funds to pay for food, which is a direct result of the persistence of food as a commodity.

When organic and fair trade or even locally produced food remains a commodity to be bought and sold, reforms to the existing system cannot resolve this contradiction. For example, social movements in the UK were early advocates for organic agriculture, and a variety of successful experiments to transform the food system emerged in England, Scotland and Wales. However, food insecurity persists in these places with up to 8 million people still struggling to eat well every day. These people include farm workers and farmers who earn less than the median income and receive less than 10% of the value of their food that is sold in markets. Researchers and advocates argue that the UK needs a "People's Food Policy" that puts the needs and well-being of eaters and producers as a policy priority for the governance of the food system (Butterly & Fitzpatrick, 2017). For them, powerful supermarkets control too much of the value of food and are supported by governance that benefits. They advocate for democratizing governance, reorganizing markets and reforming land and labor practices to disrupt the two-class food system.

FOCUS SECTION: REPRESENTATION

The alternative food movement has spawned a global culture of foodie-ism, or the glorification of gourmet interests in food. This is nowhere more apparent than the cult of celebrity chefs who coddle baby vegetables and extol the virtues of artisan butter and a generation of food writers and critics who skewer the food system for its injustices. What this culture has yet to grapple with is the overwhelming whiteness (and frequently maleness) of its representatives. From Jamie Oliver to Michael Pollan, the view we are fed is a whitened version of the world, often spawning new forms of appropriation and commodification of other cultures and their foods, from which they profit (Kong, 2019). Critics urge the publishing and media industries that support these representatives and their worldviews to dismantle the dominance of whiteness and maleness in these narratives by supporting chefs of color who are engaging with their own food traditions in ways that are less appropriating as well as providing new insights for readers and viewers into cultures and foods. Examples of such efforts in North America include Chef Yazzie, a Navajo chef who celebrates Indigenous American food and Salmon n' Bannock Bistro in Vancouver, British Columbia, the only Indigenous-owned restaurant in the city. In the UK, Ching He Huang, a Chinese-British chef, was recently granted media spots and awards, and Ravinder Bhogal's vegetarian restaurant Jikoni: No Borders Kitchen celebrates a fusion of British and South Asian cuisine.

SUMMARY

The price, value and cost of a food product, and therefore its availability in certain places and to certain people is not a function of markets alone. It is a socially constructed process that is determined by the context of the commodity and its passage through food supply chains that are governed by actors with varying degrees of power in particular ways for particular outcomes. As Kate Raworth (2017) says in her book *Doughnut Economics*, changing the goal will change the system. If we change why we value food – because it provides necessary sustenance for every person on the planet – we will change how we value it as well. Given the way power works in the global food system, often marginalizing producers and consumers, this will take a coordinated effort on the part of states, markets and civil society to resolve the contradictions of the food system that make some foods available to some people and not others. The directions of such a movement are the subject of the next chapter.

SUPPLEMENTAL MATERIALS

Key Terms and Concepts: chronic hunger, embeddedness, food desert, food insecurity, food loss, food swamp, food waste, malnutrition, neoliberalism, overnutrition, two-class food system, undernutrition

Explore More: The World Food Programme, an organization operating under the umbrella of the United Nations, works to end hunger worldwide. Their HungerMap monitors hunger globally in near real time. https://hungermap.wfp.org/

Recommended Viewing: *Supersize Me* (2004), *The Informant!* (2009)

Discussion Prompt: How do you see inequality with regard to access to food in the food system? Perhaps you have experienced it yourself? What do you feel are solutions to the problem? Aim for 100–250 words. Be sure to correctly use at least *one concept* from this chapter *or* another chapter in your discussion post and write it in **bold** type.

REFERENCES

Butterly, D., & Fitzpatrick, I. (2017). A people's food policy. *Land Workers' Alliance, Ecological Land Co-operative, Centre for Agroecology, Water and Resilience, Global Justice Now, and Permaculture Association, 24.*

Buzby, J. C., Farah-Wells, H., & Hyman, J. (2014). The estimated amount, value, and calories of postharvest food losses at the retail and consumer levels in the United States. *USDA-ERS Economic Information Bulletin*, (121).

DuPuis, E. M., & Goodman, D. (2005). Should we go "home" to eat?: Toward a reflexive politics of localism. *Journal of Rural Studies, 21*(3), 359–371.

Elton, S. (2019). Reconsidering the retail foodscape from a posthumanist and ecological determinants of health perspective: Wading out of the food swamp. *Critical Public Health, 29*(3), 370–378.

Global Nutrition Report. (2020). 2020 Global Nutrition Report. Retrieved April 11, 2022, from https://globalnutritionreport.org/reports/2020-global-nutrition-report/.

Guthman, J. (1998). Regulating meaning, appropriating nature: The codification of California organic agriculture. *Antipode, 30*(2), 135–154.

Hinrichs, C. C. (2000). Embeddedness and local food systems: Notes on two types of direct agricultural market. *Journal of Rural Studies, 16*(3), 295–303.

Kong, J. (2019). Performing authenticity and commodifying difference in celebrity chefs' food and travel television programmes. In *Routledge handbook of food in Asia*. Routledge.

McMillan, T. (2012). *The American way of eating: Undercover at Walmart, Applebee's, farm fields and the dinner table*. Simon and Schuster.

Patel, R. (2005). *Stuffed and starved: The hidden battle for the world food system*. Melville House Pub.

Raworth, K. (2017). *Doughnut economics: Seven ways to think like a 21st-century economist*. Chelsea Green Publishing.

Rose, D., Bodor, J. N., Swalm, C. M., Rice, J. C., Farley, T. A., & Hutchinson, P. L. (2009). *Deserts in New Orleans? Illustrations of urban food access and implications for policy*. University of Michigan National Poverty Center/USDA Economic Research Service Research.

Treanor, J. (2015, October 13). Half of world's wealth now in hands of 1% of population. *The Guardian*.

Walker, R. E., Keane, C. R., & Burke, J. G. (2010). Disparities and access to healthy food in the United States: A review of food deserts literature. *Health & Place, 16*(5), 876–884.

Winter, M. (2003). Embeddedness, the new food economy and defensive localism. *Journal of Rural Studies, 19*(1), 23–32.

CHAPTER 6

The Right to Food

INTRODUCTION

In October 2009, agricultural officials in the US state of Georgia raided the distribution site for the online farmers market Athens Locally Grown (ALG). Operating on a tip that the market may have been illegally selling uninspected meat, they instead found and seized (without a warrant) unpasteurized milk from a neighboring state, where it is legal to sell. The milk was impounded at the home of the founder and creator of ALG, Eric Wagoner, while federal officials were alerted to a possible violation of interstate commerce laws. Five days later, on the following Monday morning, state and federal agents met at Wagoner's home to supervise the destruction of the milk, which involved dumping the 110 gallons of milk on Wagoner's driveway (shown in Figure 6.1), although several disappointed customers disposed of the milk by drinking it. The outrage that followed resulted in the unsuccessful introduction of a bill in the Georgia legislature to legalize the sale of raw milk in the state (see also Kurtz et al., 2013).

Wagoner's case resulted in a lawsuit against the Food and Drug Administration (FDA) and the US Department of Health and Human Services, with plaintiffs in other similar cases. The plaintiffs argued that the FDA ban on interstate commerce in unpasteurized milk infringes on rights to travel and to privacy and abrogates substantive due process. After 2 years in which the FDA tried repeatedly to have the suit thrown out, the case was dismissed by a federal judge in March 2012 who cited that there is not a *universal right to consume the food of one's choosing*. The state's right to govern in the way that it sees fit, whether in the interests of the public or not, trumps the right to food. The campaign to legalize raw milk in Georgia drew attention to the fact that public health departments work with the dairy industry to protect big agribusinesses by restricting access to alternatives. While unsuccessful, the campaign did not prevent the sale of raw milk (only the consumption by humans). In a bizarre compromise that fools no one, it is only legal to sell raw milk in Georgia to be consumed by pets.

This chapter extends the insights from the previous section to discuss social and political movements to expand access to food that reveal how the right to food is not a guarantee, and in some cases, governance re-entrenches colonial and capitalist systems of power. Food aid, food justice and food sovereignty approaches are discussed to show how food insecurity and lack of access to food are not accidents of fate but are produced through place-specific systems of uneven development. The chapter concludes with a discussion of

DOI: 10.4324/9781003159438-8

FIGURE 6.1 Raw milk dumping in Georgia (photo credit: Edwyna Arey)

sustainable development and captures the debates generated by the successes and failures of these movements and their deep relationship to place.

KEY THEMES AND IDEAS: EXPANDING FOOD ACCESS

The **right to food** is enshrined in the 1948 charter of the United Nations (UN): "Everyone has the right to a standard of living adequate for the health and well-being of himself and of his family, including food, clothing, housing and medical care and necessary social services, and the right to security in the event of unemployment, sickness, disability, widowhood, old age or other lack of livelihood in circumstances beyond his control" (Article 25). This right currently derives from the International Covenant on Economic, Social and Cultural Rights to which 170 states have signed as of April 2020. This covenant commits states to ensuring food security. Food security is defined by the UN as "a situation that exists when all people, at all times, have physical, social and economic access to sufficient, safe and nutritious food that meets their dietary needs and food preferences for an active and healthy life" (IFPRI, 2021). The map in Figure 6.2 indicates which countries have signed the covenant and where they are in the process of developing state-based laws to ensure food security. The US, as indicated previously, has no known right to food, although the US state of Maine passed a right to food law in 2021.

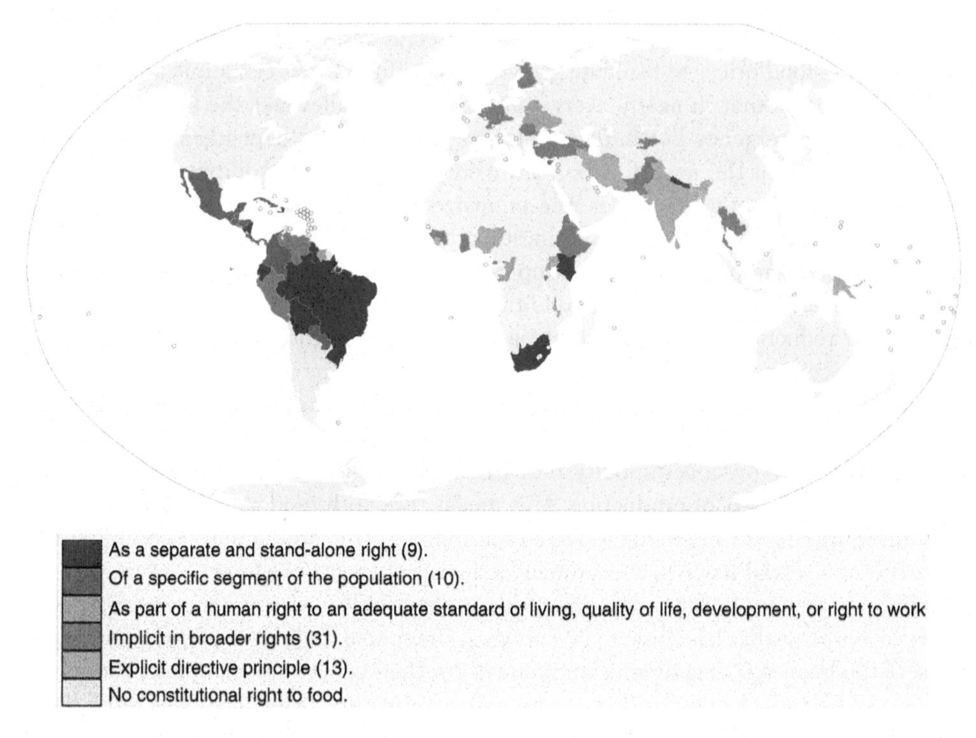

FIGURE 6.2 Right to food (https://commons.wikimedia.org/wiki/File:Map_of_constitutional_right_to_food.svg)

FOOD AID PROGRAMS

Food aid is food that is given out for free in an emergency situation to prevent starvation. Food aid has domestic and international foci and are often combinations of states, markets and civil society working together. State-run food security programs, premised on the notion that people should have access to safe, adequate, and appropriate food, emerged with the development of the welfare state in the 1960s, primarily in more developed countries. In the US, food security policies responded to both the over-production of commodities and widespread poverty during the Great Depression and sought to ensure access to food through a variety of state-based mechanisms, including credits that allocate a certain amount of money per qualifying household toward the purchase of food. This model has since expanded to many more nation-states, particularly during the Cold War era.

Additionally, the technological changes brought to bear on agriculture via the Green Revolution in developing countries were an exercise in **philanthrocapitalism** – the giving of aid in order to cultivate new markets – and were justified by their ostensible goal of mitigating food insecurity. The development of policies that employ market mechanisms to distribute food to the poor are consistent with microeconomic policies and neoliberal notions of the subject that make the individual responsible for nutritional intake via the purchase of food or the receipt of it as food aid. Subsidies for commodities produced in the developed world also produce surplus to be used as a tool of foreign policy and artificially suppress food prices to facilitate growth and profit in other economic sectors.

Research shows that in nearly every context, food aid alleviates the short-term need for food in an emergency but initiates a long-term pattern of dependence because the underlying causes of the need for food aid were not addressed. Additionally, the global circulation of commodities such as rice or maize reduces local prices, lowers farmers' incomes, and ultimately undermines domestic production. Both the depression in income and the loss of domestic production set up conditions for dependency on foreign sources of food. Food aid often comes with conditionalities, such as accepting genetically modified seeds or reductions in spending on health or education. Additionally, policies such as subsidies for commodity production encourage oversupply in countries from which the aid comes, which creates systemic dependence and poverty. Export models of agricultural development, such as the production of coffee, also produce situations of vulnerability in which farmers grow a commodity they can neither consume nor sell if global prices decline below the costs of production. This means that both food aid (the production of consumers) and export-oriented models of development (the production of commodities) are part of the **spatial fix.** When a company exhausts either its productive resources or its consumer market (or both) in one place, executives seek new resources or markets elsewhere in a process that has come to be known as the spatial fix (Harvey, 2003). For example, seed companies, seeing diminishing returns for their genetically modified products in the US and being locked out of Europe by policy, turned to Africa. They marketed seeds to farmers and developed test plots while they cultivated markets in the region.

FOCUS SECTION: DIRECT PAYMENTS

A promising and productive solution to food insecurity recently has been the trialing of direct cash payments to households. The cash does not necessarily require the household to purchase food, and it may be used to capital development projects, such as the purchase of livestock or farmland, which can reduce the vulnerability of the household to food insecurity (Miller et al., 2011). When the cash is used to purchase food, it is far more likely to be used to buy food from local farmers, thereby stimulating a local economy and reducing the vulnerability of other farmers as well. Universal basic income experiments also operate on this same principle, and recent studies indicate that providing for people when they need help is cheaper in the long run than addressing the health and other social problems that poverty and chronic hunger produce (Bregman, 2017). In short, helping people by giving them money, especially when it is not restricted or conditional on job-seeking for example, makes the most sense in a system that operates on money.

CONCEPTUAL ENGAGEMENT: SUSTAINABILITY

In 1987, the Brundtland Report, written by the World Commission on Environment and Development in response to a request for an agenda for change from the UN, defined

sustainable development as "the idea that human societies must live and meet their needs without compromising the ability of future generations to meet their own needs" (World Commission and Brundtland, 1987). This call for change came in the wake of environmental disasters and the looming threat of climate change. In theory, this meant that capitalism would continue uninterrupted but that companies would be encouraged to focus on the **triple bottom line** instead of the single bottom line of profitability. The triple bottom line includes social equity, economic profitability and environmental soundness. In practice, this resulted in business as usual with a lot of "greenwashing," such as petroleum companies calling themselves sustainable when they developed solar technology, while increasing oil production at the same time.

In 2000, the UN established the Millennium Declaration Goals, aimed at reducing poverty and improving health outcomes and ecological health by 2015. They were replaced by the Sustainable Development Goals in 2015 to be met in 2030, because it was clear that the earlier, similar goals would not be met in time. This seemingly uphill battle for global well-being confronts the persistence of capitalism globally, which profits from inequality and the unsustainable use of environmental resources. The social movements to end hunger and poverty, such as food sovereignty, advocate for change in the economic systems that produce hunger and inequality and identify place-specific and local solutions to global problems. This echoes earlier calls for decentralizing solutions to environmental problems in the 1992 UN document Agenda 21. Nation-states rejected such calls for local control, arguing that it would undermine state sovereignty. And now, years after the Sustainable Development Goals were put in place, hunger continues to grow.

FOCUS SECTION: BLACK PANTHER AND NATION OF ISLAM

The Black Panthers, a political group organized around liberation and justice for African Americans, viewed hunger as a violent form of oppression. They worked to end this for impoverished children by providing free meals in schools in the 1960s and 1970s. Beginning with breakfast, they grew the program to three meals day. Many working families, due to the suppression of wages, which were often discriminatorily low for Black workers, could not afford to buy food. The Black Panthers procured food for the program through donations from local businesses, and the success of their organizing paved the way for the federal free and reduced breakfast and lunch programs in US schools today. The Nation of Islam in rural southern Georgia has a similar mandate, aimed at feeding all Black people in America. Through the space of Muhammad Farms, they celebrate and reframe Black agrarian history in the US and focus on transforming the farming landscape in a space of liberation and transformation through a return of African Americans to the land (McCutcheon, 2013).

Sustainable food systems are difficult to measure, but one research group developed a metric that incorporates 20 indicators of sustainability under four general dimensions: ecological, economic, social and food security and nutrition (Béné et al., 2019). The indicators include biodiversity, gender equity, food safety, food security, food waste, wildlife and energy, among many others. Countries scoring high include Canada and the UK, while countries scoring low include India and Bangladesh. Two observations can be made from these patterns that are consistent with the food system in general: the wealthier countries that were former colonizers or settler states of empires generally score higher, presumably because of the high overall nutrition and health of the population and the availability of resources for agriculture, both of which are a function of imperialism. The second observation follows from the first: measuring sustainability is extremely difficult and will also follow the values that are socially constructed to contribute to sustainability. For example, the percentage of farms engaged in direct sales was not an indicator of sustainability, while many scholars and activists would suggest it is. If it was, the countries that currently rank low would shift higher as would many of the countries ranked high, would shift lower.

In the example that opened this chapter, unpasteurized milk is declared illegal and unsafe to consume in some places but not others and is not regulated in many of the countries thought to have unsustainable food systems. Some countries have laws that forbid the consumption of raw milk, some of whom border countries that do in Europe, so it is clearly not an environmental issue. For example, the cows in Figure 6.3 are famous for their milk that is used in celebrated raw milk cheeses. While it may be a result of varying governance of production methods that are deemed unsafe, it is clear that raw milk itself is

FIGURE 6.3 Montbéliarde cows in France

not inherently dangerous. What does explain the variation is the existence of pasteurizing equipment, in the US, Canada, Australia and Europe, where raw milk is regulated. This technology effectively creates a bottleneck in the supply chain at the point of processing. Recall from the simplified food system diagram in Chapter 1 that there are many producers and consumers and few processors, and that value is added at each link in the supply chain. Therefore, power is concentrated in a few facilities that control pasteurization in the milk supply chain, and their power is enforced through state law. The justification for such concentration is allegedly food safety and the public interest, but it is belied by the instances of contamination that occur far more often in the pasteurized milk supply chain than in the short and local supply chains that often characterize legal (and illegal) raw milk sales.

This nation-state's ability to control what can be consumed is often referred to as biopower. **Biopower** is the capacity to manage the health of human populations with "population health" as a political object (Foucault, 2003). Foucault (2003) writes that in the 18th century, a particular kind of medicine emerged whose primary purpose was to ensure public health through hygiene, with institutions, such as the state working to ensure worker productivity and health. States exert biopower by managing population health in arenas including public hygiene, clean water supplies, food safety and nutrition. The growth of biopower since the 17th century entailed an alignment between the health sciences and technologies of hygiene, the administrative capacity of the state, and the dictates of capitalism. Food safety law is one such realm of state power that dictates how, where and what we eat. Such laws aim to protect us from contaminated food sources, but they also do not prevent contamination that is produced through unjust labor practices, environmental pollution and supply chain failures, the examples of which are countless. They also prevent the consumption of foods of one's choosing, such as raw milk mentioned earlier.

FOCUS SECTION: ETHICAL DIETS

Ethical diets consume much of the attention of a public concerned with distributing the goods and bads of the food system fairly. One such movement is a push to consume fewer animals as an antidote to climate change as well as to address animal welfare and working conditions in processing plants. Meat consumption increases as a society becomes wealthier, and while it is widely agreed that Americans, in particular, consume more than their fair share of meat and thus disproportionately contribute to the "bads" of food production, experts do not agree that halting meat consumption altogether is necessary or wise. The reasons for this are multiple and complex, but at its most basic level, protein replacements, such as soy, are often sourced from long supply chains, are grown in monocultures and are not carbon neutral. Livestock, such as chickens, goats and bison, which can be raised in nearly every ecosystem, contribute to small farmers' economic sustainability and can be raised in carbon-neutral ways in sustainable, local systems, particularly because they contribute to the growth of grasslands and ecosystem health. Even if one does not consume meat, milk and eggs from local farms can contribute to sustainable ecosystems and community well-being. This does not address the fact that animals must die in order to be consumed, but the perpetuation of animal breeds and entire species depend on our agricultural systems. In short, rather than eliminating them altogether, there are better ways to integrate animals into our food systems if only consumers demand them and the state allows them.

EXAMPLES: FOOD JUSTICE AND FOOD SOVEREIGNTY

Food Justice

Food justice has roots in the environmental justice movement. This movement identified that communities of color disproportionately bear the consequences of environmental pollution and sought to mitigate the harm to communities of color. This movement extended from protecting the places where people live, work and play, to what and how they eat. Food justice is defined as "ensuring that the benefits and risks of where, what, and how food is grown and produced, transported and distributed, and accessed and eaten are shared fairly" (Gottlieb & Joshi, 2010, p. 6). Food injustices are created by systemic racism and manifest in the production of food deserts, or a lack of access to fresh food in communities of color (Reese, 2019). Food injustices also include an ongoing lack of access to productive resources for Black and Indigenous farmers and inequities in the labor process that exploit people of color in the two-class food system.

Sharing roots with community-based food security, food justice activism seeks to decommodify food through community garden projects and food redistribution efforts. It often, but not always, includes charity as an often problematic approach to food injustice. Charity models are consistent with neoliberal food policies because the distributed foods are often surplus commodities, and their redistribution does not challenge the current economic structures of the food system that produce hunger and surplus in the first place. Similarly, food justice projects originating outside the community often discipline individuals to be better consumers (i.e., eat healthier food or be more educated about their food choices). Additionally, very few, if any, forms of food justice activism specifically target neoliberal policies, and they thus fail to engage with the state-based policies that develop and promote markets for food. Those projects that are led by community members, rather than outsiders seeking to do good, are generally the most successful and build self-reliance rather than dependence.

FOCUS SECTION: FOOD JUSTICE IN ENGLAND

Researchers in the UK examined the contributions of charity-led community food initiatives that responded to the rising food insecurity in England (Kneafsey et al., 2017). They examined two prominent programs of the Garden Organic project that focus on educating and empowering people to grow their own food in various spaces in the community, such as in backyards, at community gardens or at schools. They found that while the organization could not fill the gaps in access to food, the attainment of new skills and awareness about the growing of food had transformative potential. They found, in particular, that the connections that people made while working together fostered community resilience. The study also found that the potential exists for a more "radical" form of awareness that could organize people around problems in the food system and the right to food that is undermined by capitalism. So, while food justice movements cannot necessarily feed people adequately, they do contribute to a growing awareness of the sources of food insecurity and connect people around solutions.

Food Sovereignty

A significant and growing challenge to the influence of capitalism on food production, trade and regulation of the food system is a global movement operating under the banner of **food sovereignty**. Activists for food sovereignty position themselves against the corporate food regime in order to expand the meaning of human rights to include a "Right to Food" as guaranteed by the UN Declaration on Human Rights. Food sovereignty confronts what Malik Yakini of the Detroit Black Community Food Security Network calls the "twin evils of white supremacy and capitalism" (Yakini, 2013). This is a global movement led by peasant farmers and Indigenous people, who argue that food aid creates more problems than it solves and call for a restructuring of the political, cultural and economic systems that produce food. They aim to undermine corporate control of the food system through new forms of democratic decision-making at a local level. This requires the engagement of individuals in a particular place to work together to establish a system of local agriculture that can feed a community (Daigle, 2019). There are multiple roadblocks at various scales to this project, chief among them, the cheap availability of food from around the globe, loss of traditional knowledge and policies that restrict local agriculture and prevent the exchange of certain foods, such as raw milk. Central planks of the food sovereignty platform, as articulated by the delegates to the 2007 global forum on food sovereignty in Nyéléni, Mali, include local markets, **agroecology**, Indigenous rights and power over territory (Nyéléni, 2007).

LOCAL MARKETS

Food sovereignty movements do not reject markets as a tool for facilitating the exchange of food, but they demand the removal of global governance of trade and production. In the Nyéléni delegates' view, free trade policies have destroyed livelihoods and local economies. Neoliberalism, rather than guaranteeing the right to food, is in fact the source of food insecurity. The delegates recommend returning democratic control of food distribution to producers and consumers. They identify local food production, food cooperatives, local processing and **solidarity economies** or those based on mutualism rather than capitalism, as key mechanisms toward this end. In short, rescaling, decentralizing and democratizing decision-making about markets, and the development of state-based policies that support them are required for autonomy.

AGROECOLOGY

Food sovereignty is just an idea if it is not accompanied by agroecological practices that protect the resources upon which peasants and farmers depend. Theme 7 of the Declaration of Nyéléni identifies alternative production models as key to accomplishing the objectives of food sovereignty. They identify the corporate and industrial model of agriculture as ecologically destructive and seek to replace it with one based on cooperation and solidarity between people and the return to Indigenous knowledge systems. While acknowledging that ideals for the sizes for farms and for democratic participation will vary from place to place, they identify the privatization of the commons, patriarchal control of production and laws that discriminate against small-scale farmers as the largest obstacles to realizing a more ecologically sensitive agriculture. The sign on the community garden in Lisbon, Portugal, shown in Figure 6.4 requires agroecological methods to be used.

FIGURE 6.4 *Horta popular* (community garden) in Lisbon, Portugal

Indigenous Land

The Declaration of Nyéléni identifies access to land as a key right required for food sovereignty, as well as recognizing that land is a resource that is unevenly distributed through discrimination against the poor, women and Indigenous people. Agrarian reform is needed that upholds traditional and customary rights of peoples and communities so that they may have access to communal resources that were privatized through colonization. For the delegates, this also includes access to and control over Indigenous seed varieties and markets that are under community control. They also seek to ensure women's access to land and work toward the recognition of Indigenous peoples as key actors in deciding control over all productive resources, including water, land, and seed.

Territory and Power

The Declaration of Nyéléni criticizes modernist political-economic orientations toward territory that produce unequal social relations, and they are clear about the way their understanding of space and power departs from the geographic imaginary of the nation-state. They acknowledge traditional forms of knowledge, land use and land access rather than the rules of the nation-state. They stress the sharing of land over owning or privatizing territory, and they resist all forms of expulsion, whether of Indigenous people or migrants. They argue that conflicts over territories are generated through **privatization** and liberalization of land markets that leave some with much and many with too little. The key to making all of this possible is, they conclude, "decision-making power at the local level" (60).

FOCUS SECTION: WHITE EARTH LAND RECOVERY PROJECT

Broken promises litter the history of the colonial contact with the First Nations of the US, and perhaps none more so than for the White Earth Nation of the Anishinaabe people of what is now known as Minnesota. Approximately 90% of the White Earth reservation is White-owned, a spectacular land grab facilitated by a variety of treaty violations, land allotments to non-Indians and outright theft. Winona LaDuke, an Anishinaabe woman, politician and activist, started an organization called the White Earth Land Recovery Project to advocate for the return of lands to the White Earth Nation. Wild rice, a native grain endemic to the upper Midwest and Southern Canada, is harvested from ponds on the reservation and is sold by the non-profit to help sustain the organization's activities, which also include a Farm to School program in the tribal schools. The return of land to the collective management of the White Earth Nation will mean returning access to lakes and rivers where wild rice grows. Access to the lakes where it grows is vital not only to the physical health of the tribes' members but also to the dignity of the culture and the social-economy of the tribe. The food sovereignty movement on the White Earth reservation thus creates a space, in the ricing ponds, that brings together the political, cultural, economic and environmental dimensions that sustain the social life of the tribe.

SUMMARY

Inequitable access to food is a global problem. States have been tasked with solving it ever since the concerns about population health as a result of undernutrition presented challenges to society and the economy. Various efforts, including giving away food or relying on civil society, have been patches on a broken system that relies on the commodification of something vital to human health. The social movement of food sovereignty suggests a retooling of the system is required – one that changes the power relationships between people, states, land and food. Advocates for food sovereignty argue that because power is so concentrated in the hands of private corporations, the push to profit from food overwhelms the social need for sustenance and nutrition. Movement actors urge nation-states to resist the influence of corporations and capitalism in the interest of feeding people and providing food security and at the same time building more sustainable and resilient food systems.

SUPPLEMENTAL MATERIALS

Key Terms and Concepts: agroecology, biopower, food aid, food justice, food sovereignty, philanthrocapitalism, privatization, right to food, solidarity economies, spatial fix, sustainable development, triple bottom line

Explore More: The Barilla Centre for Food and Nutrition Foundation built an interactive map showing food sustainability scores for 67 countries. Explore more here: https://foodsustainability.eiu.com/heat-map/

Recommended Viewing: *Gather* (2020), *Farmageddon* (2011)

Discussion Prompt: How do you see autonomy with regard to access to food in the food system? Perhaps you grow your own food or have in the past? What do you feel are solutions to the tensions between governance and freedom? Aim for 100–250 words. Be sure to correctly use at least *one concept* from this chapter *or* another chapter in your discussion post and write it in **bold** type.

Recipe: Wild rice is a staple food for Indigenous people in the Upper Midwestern US and Southern Canada. Native Harvest, a non-profit based in the US state of Minnesota on the White Earth Indian Reservation sells it to the general public. You can find more information and recipes here: https://nativeharvest.com/blogs/news/native-harvest-wild-rice-recipes

REFERENCES

Béné, C., Prager, S. D., Achicanoy, H. A., Toro, P. A., Lamotte, L., Bonilla, C., & Mapes, B. R. (2019). Global map and indicators of food system sustainability. *Scientific Data*, 6(1), 1–15.

Bregman, R. (2017). *Utopia for realists: And how we can get there*. Bloomsbury Publishing.

Daigle, M. (2019). Tracing the terrain of Indigenous food sovereignties. *The Journal of Peasant Studies, 46*(2), 297–315.

Foucault, M. (2003). *Society must be defended: Lectures at the College de France, 1975–76* (M. Bertani & A. Fontana, Eds., English series A. Davidson, Ed. & D. Macey, Trans.). Picador.

Gottlieb, R., & Joshi, A. (2010). *Food justice*. MIT Press.

Harvey, D. (2003). *The new imperialism*. Oxford University Press.

IFPRI. (2021). *Food security*. Retrieved April 19, 2021, from www.ifpri.org.

Kneafsey, M., Owen, L., Bos, E., Broughton, K., & Lennartsson, M. (2017). Capacity building for food justice in England: The contribution of charity-led community food initiatives. *Local Environment, 22*(5), 621–634.

Kurtz, H., Trauger, A., & Passidomo, C. (2013). The contested terrain of biological citizenship in the seizure of raw milk in Athens, Georgia. *Geoforum, 48*, 136–144.

McCutcheon, P. (2013). "Returning home to our rightful place": The nation of Islam and Muhammad farms. *Geoforum, 49*, 61–70.

Miller, C. M., Tsoka, M., & Reichert, K. (2011). The impact of the Social Cash Transfer Scheme on food security in Malawi. *Food Policy, 36*(2), 230–238.

Nyéléni. (2007, February 23–27). *Forum for food sovereignty*. Proceedings held in Sélingué, Mali. https://ddd.uab.cat/pub/estudis/2007/180504/nyelenien_a2008.pdf?msclkid=6597f9e7aa8611ecb00d214773cfd729.

Reese, A. M. (2019). *Black food geographies: Race, self-reliance, and food access in Washington, DC*. UNC Press Books.

World Commission on Environment and Development, & Brundtland, G. H. (1987). *Presentation of the report of the world commission on environment and development to the commission of the European communities, the EC and EFTA countries . . . 5 May 1987, Brussels*. World Commission on Environment and Development.

Yakini, M. (2013, September). Address at Food Sovereignty: A Critical Dialogue, Yale University. *YouTube*. Retrieved April 11, 2022, from https:// www.youtube.com/watch?v=_LaMt9HVQFY.

PART 3

What We Eat

CHAPTER 7

Produce

Fruits and Vegetables

INTRODUCTION

Some of my earliest memories are of picking wild blueberries in the northern pine woods of my childhood home in the US state of Minnesota. Blueberries, like most fruits, are nutritional powerhouses, with a variety of vital nutrients key to human health, including anthocyanins, powerful antioxidants found in the skins of blueberries. Like blueberries, most of today's fresh fruits do not stray too far from their wild ancestors, and Indigenous people likely practiced pre-domestication cultivation with many of the fruits we consume today. They selectively adapted the environment to favor the foods they preferred and built entire ecosystems around the plants that helped them live good lives. Michael Pollan (2002) writes in *The Botany of Desire* that plants chose us for domestication by appealing to our desires for sweetness, intoxication and beauty. The wild blueberries shown in Figure 7.1 are in an alpine meadow, and their sweetness, bright blue color and foliage no doubt attracted early hunter-gatherers who likely cultivated prime habitats for them.

This chapter begins a section of the book focusing on different types of commodities, their supply chains and contemporary issues associated with their production and consumption. This chapter includes summaries of broad categories of fresh fruits and vegetables, and their places of production and the (neo)colonial relationships, labor relations and technological innovations that continue to shape supply chains. It will elaborate on a thread initiated in Chapter 1 on potatoes and will include discussion of the Banana Trade War, jackfruit as a tool of resistance and super foods.

KEY CONCEPTS AND TERMS: FRUITS AND VEGETABLES

Botanically, a **fruit** bears the seed of any flowering plant and is the mechanism of seed dispersal by angiosperms. It is believed that edible fruit-bearing plants co-evolved with humans and other animals in a symbiotic relationship. Plants use fruits to tempt us, birds and other animals to eat them as a strategy for seed dispersal. Animals received nutrition; angiosperms had their seeds distributed. Fruits are now a significant part of the global primary sector and are sourced from every region in the world. Some fruits, such as apples and pomegranates, have taken on specific cultural meanings in myth and story. For example,

DOI: 10.4324/9781003159438-10

FIGURE 7.1 Wild blueberries (https://commons.wikimedia.org/wiki/File:Blueberry_In_The_Mountains_(248541869).jpeg)

the Greek legend of Persephone tells the story of a world with an eternal summer until Hades, the god of the Underworld, kidnapped her. While there, she ate six pomegranate seeds before she was released, thus ensuring her return for 6 months of the year, during which winter would prevail in the world above. This tells a story of the natural seasons of dormancy and renewal that are central to fruits. As **perennials**, or plants that do not need to be grown from seed every year, most fruit-bearing trees require a certain number of days below freezing to guarantee blooms. Not all fruits are sweet, however, and many of the plants we categorize as vegetables are actually the seed-bearing anatomy of the plant, but are savory or starchy, including beans and grains.

Any edible part of a plant is considered a **vegetable**, including the flowers, fruits, stems, leaves, roots and seeds. For some plants, like beets or celeriac, the whole plant can be consumed. Obviously, there is considerable overlap between fruits and vegetables and even grains, but vegetables entered human diets through gathering practices that eventually led to domestication. Over time as trade developed between places and regions, more and more plant crops became integrated into the human diet. Vegetables are grown and consumed all over the world with place-specific cuisines attached to them. Because of the wide variety of plants that are consumed in the world, and their relative ease in cultivation, vegetable production ranges from backyard gardens to hundreds of acres of monocrops harvested with machinery. As **annuals**, or plants that generally require seeding each year, production includes planting, transplanting, cultivating, harvesting, grading, processing, storage and marketing. Depending on the type of vegetable concerned, harvesting the crop is followed by grading, storing, processing and marketing. Nutritionists recommend making fruits and vegetables the basis of any diet due to the high levels of phytonutrients including fiber, vitamins and minerals. Vegetables have slightly more variation in form than fruits, and some have lower perishability and can be easily stored. Vegetables at high latitudes require greenhouses and other amendments to be grown in the winter.

FOCUS SECTION: THE GLOBAL POTATO

Potatoes, originating in the Andes of South America and transported around the world through colonization, now appear in nearly every cuisine. As an easily produced, highly palatable source of carbohydrates, potatoes have been adapted to cuisines everywhere in place-specific ways. In France, the predecessor of french fries, *pommes frites* are popular with meat dishes and as snack food. In Eastern Europe, potatoes are shredded and fried and associated with religious observances. In pasta-loving Italy, potatoes are mixed with semolina flour and boiled to make gnocchi. In England, potatoes top a baked meat, often lamb, dish. In Canada, formerly a French colony, *pomme frites* are doused with cheese and gravy to compose the national dish of poutine. In Ethiopia, long a trading partner with India, potatoes are cooked with cabbage and carrots and fragrant spices from South Asia. In Malaysia they are cooked like a curry with onions and fiery spices. In India, fried mashed potatoes are served as street food, often as a patty in a bun. In Sri Lanka, potatoes are cooked with jackfruit in a popular curry. There is seemingly no end to the possibilities for the versatile global potato.

In spite of the extreme seasonality of fruits and vegetables, consumers in the most developed economies have come to expect year-round abundance in the supermarket. The gleaming rows of produce never change, no matter what the season. The reason people in North America and Europe can expect to see eggplant and strawberries every day is due to long-established patterns of trade, many of them colonial in origin. The bulk of fruits from Central America are sold into US markets, while South American produce is sent to Europe and Asia. South Africa, a former Dutch colony, sells seasonal produce into European markets. With global trade comes technological innovation to help highly perishable fruits survive transportation over long distances, as well as labor abuses and environmental contamination in places that are often, but not always, very far away and thus out of the conscious mind of many consumers.

Fresh fruits are extremely seasonal, very perishable at their peak and must be preserved in some form through drying, freezing and canning. Preservation takes a variety of forms and can include drying, pickling, fermenting, jellying and freezing. Given that fruits and some vegetables are filled with sugar and often energy packets in the form of seeds, they are vulnerable to rapid decomposition from microorganisms, such as bacteria and yeasts. Some forms of preservation, such as jellying, work to kill the bacteria through heating, preventing the growth of more bacteria by sugaring and preventing recontamination through sealing in airtight containers. Other methods use the presence of bacteria and yeasts, such as wines or vinegars, and preserve fruits through fermentation, or encouraging the growth of yeast, which produces alcohol or acid and prevents the growth of microorganisms. Lacto-fermentation is a form of fermentation used to preserve vegetables, such as cabbage to make sauerkraut or kimchi. This method uses salt to kill pathogenic bacteria, which cannot tolerate salt, and further the growth of helpful bacteria, which can tolerate salt. The ability to preserve food has, until relatively recently, been a source of food security for people living in highly seasonal, high latitudes. Preservation has largely fallen out of practice, but it reduces waste and has the capacity to contribute to community-scale food security and improve access to nutritious food for the most vulnerable. Community canneries are an idea whose time has come and which are beginning to be integrated into community garden projects.

FOCUS SECTION: SUPERFOODS

Superfoods are one of the newest rages in the world of food fads. Demand for such products as acai berries and avocados, which have their origins in Latin America, is surging, largely due to the influence of social media. Consumers increasingly look beyond labels for nutrition and health information about the foods they consume to the internet, resulting in the perpetuation of false or misleading claims (Butterworth et al., 2020). In the case of the so-called superfoods, claims about the ability of a given food to provide a given health benefit are not likely to be supported by scientific research and are used by marketers to increase sales. Being seen on social media eating such foods, often by self-styled influencers, has also contributed to the communication of dubious health-food claims. Like many foods sourced from tropical regions, the export of food, whether "super" or not, often results in an increase in food insecurity and dependence for the farmers who produce it. In addition, the consumption of superfoods often follows the pattern of unequal access to food, with people of higher socio-economic class more likely to consume them.

CONCEPTUAL ENGAGEMENT: CONTEMPORARY PRODUCTION

The production of fruits and vegetables has been transformed from the kitchen gardens of the earlier 20th century to the large-scale industrialized monocultures of the early 21st century. Recent technological innovations in fruit and vegetable production include genetic interventions, chemical inputs, robotic harvesting and the use of artificial intelligence to produce commodities. Genetic interventions include making them more resilient to mechanical harvesting and transport in long supply chains, which often leave us with woody tomatoes and watery strawberries. They also include failed biotech attempts at making fruits and vegetables immune to freezing. These interventions simplify the genetic diversity of the crop, requiring inputs to fight disease and pests. Labor shortages, unionizing efforts to fight exploitation and consumer push for more ethical labor practices have pushed some parts of the industry into more mechanical harvesting. While still requiring human labor, the number of people required is far less than if fields were to be harvested by hand. Artificial intelligence is also used to identify pests and to target mitigation, which is an alternative to blanket spraying of entire fields. To afford the capital investments that such technology requires, however, producers must scale up production, which if you recall from the diagram of agricultural modernization has a positive feedback effect, requiring more land, more technological innovation, fewer people, greater urbanization and ultimately a greater demand for industrialized food production.

Labor is an ongoing challenge in agriculture, which is discussed in much more detail in a separate chapter. Because food has a relatively low **profit margin** (e.g., the profit on a pound of tomatoes is small compared to the profit on an electronic), when all other input costs are relatively fixed (land, inputs, seeds), labor becomes a place of perceived or necessary flexibility in the production process. Producers will thus recruit a vulnerable and very often discriminated against class of people to work in agricultural fields. In the Dominican Republic banana plantations, these workers are Haitians who live in the Dominican Republic temporarily or illegally. In Europe, these are migrants from former Soviet Republics in Eastern Europe. In Israel, workers migrate from Thailand. The list goes on. Migrants, as non-citizens in the country within which they work, have fewer rights and are vulnerable to deportation if they request such rights. In addition, in the US, agricultural labor, ostensibly due to its seasonal nature and also family farm histories, there are several key exemptions to labor laws that govern other kinds of work: child labor is allowed, there is no minimum wage and overtime is not compensated at a different rate. This means that people under the age of 16 years may work in fields for a **piece rate** (a wage based on how much they pick) and have no cap on how many hours they can legally be allowed to work. Add to this, that this is back-breaking work undertaken in conditions that are often hot and contaminated by chemicals (McMillan, 2012). Workers shown in Figure 7.2 pick tomatoes on an organic farm in Pennsylvania. The Coalition of Immokalee Workers estimate that a worker being paid by the piece (in this case a crate) will have to pick more than 2 tons of tomatoes to earn a minimum wage in one day of work.

In the 1960s, Rachel Carson exposed the dangers to human and environmental health of agricultural chemicals, including, and most famously, DDT, which disrupted the reproductive cycles of birds and is considered a possible human **carcinogen**, or cancer-causing compound. Its use has been restricted in the US since the 1970s, but there are many more examples of agricultural chemicals that pose enormous risks to human and environmental health. Strawberries and the various chemicals used in their production form a current and ongoing battle documented by geographer Julie Guthman (2017). Strawberries, as sugar-filled packets growing close to the ground, and which must be picked just before the peak of ripeness, are vulnerable to a wide array of pests, fungi and diseases. The chemicals used to combat them include fungicides and pesticides that have been key to the industry's success, but they are highly toxic to the workers who pick the strawberries and get strawberries on the list of the **dirty dozen** fruits and vegetables that are laden with chemicals at consumption. Consumer efforts to restrict the use of chemicals sparked a public debate about the hazards and necessity of their use. Activists, advocating largely for consumers, argued, sometimes successfully, for bans on their use, while industry argued that the industry would collapse if they were banned. Some chemicals were ultimately phased out, leaving the strawberry industry precarious due to its own reliance on toxic chemicals,

FIGURE 7.2 Farm workers in a Pennsylvania tomato field

making it "a cautionary tale about industrial agriculture . . . scaled-up, simplified **monoculture** [bold added] accompanied by forms of exploited and often spatially transported labor" (Guthman, 2017, p. 10).

The Alar apple scare in the 1980s alerted consumers to carcinogens in the produce supply chain and led to public outcry and an increased demand for certified organic food. Alar is a ripening agent that is sprayed on apples to make them ripen on the tree before they fall off. It breaks down into a carcinogenic compound, which exposure to is estimated to cause 100 cancers in 1 million people, 100 times the acceptable threshold for harm in human populations. This was confirmed in lab tests with rats in 1973, 1977 and 1978. In spite of the evidence, its use continued for several years, even though the compound was detected in apple products, including those developed for infants and children. In 1980, the US Environmental Protection Agency opened an investigation into the hazards associated with its use but closed it after meeting with Alar's manufacturers. Soon after, Massachusetts and New York, large apple growing states, banned its use, and the American Academy of Pediatrics endorsed a federal ban. In 1989, the television program *60 Minutes* aired an exposé, the Natural Resources Defense Council enlisted Meryl Streep in an ad campaign against its use, citing the grave risk of cancer, particularly among children. Alar was dropped from use that same year. This example shows the power of public opinion to influence production practices, and the relative weak governance by regulatory agencies at the national-state scale.

As **food scares** or events that call attention to the danger in a particular food item, such as the Alar scare, became more common in the late 20th century, consumers began to experience deep anxieties about the origins of their food, particularly in the long, illegible supply chains that had come to dominate out-of-season and tropical commodities. The places of production of tropical fruits and vegetables are places of export, not import, and they are also not usually places of consumption for valuable exports. Figure 7.3 shows a cashew fruit, which bears a single cashew that must be harvested and processed by hand. The Ivory Coast and India are leading producers and exporters and have increased production to meet the rapid increase in demand for nut-based milk in the Global North.

Geographer Susanne Freidberg (2004) in her book *French Beans and Food Scares*, writes about the **governance** of supply chains. She details the development of systems of auditing, quality control, standards of hygiene and traceability in the supply chain for fresh produce to Europe from former colonies in Africa. In the case of French beans, British consumer demand for quality and purity imposed a set of technological, moral and administrative demands on production and distribution that British exporters dominated, controlling the supply chain in neocolonial ways. She writes, "the power to demand goodness in food – as defined as cultural norms of what makes food safe, natural, moral and appetizing – has introduced new forms of domination and vulnerability" (5). Processing occurs

FIGURE 7.3 Cashew fruit

in some cases in **special economic zones** where labor regulations and taxes and tariffs are often suspended for the exporter. Recall from the simplified food system/supply chain in Chapter 1, that wherever there are few actors, power and capital will be concentrated. This example also demonstrates how the global food system replicates the imperial system of sourcing food from former colonies to meet the needs of colonizers. Recall from chapter 1 that all food systems are governed by three main actors: states, markets and civil society. When powerful states come into conflict with markets, the outcomes can be less than certain and often result in protracted negotiations, such as those seen in the Banana Trade War.

EXAMPLE: THE CASE OF THE BANANA TRADE WAR

Bananas, one of the most ubiquitous and banal of tropical fruits, belies a violent colonial history and ongoing struggle over what is produced, where and under what conditions. What follows is an outline and description of what led up to the conflict known as the "Banana Trade War" between the European Union and the US in the late 20th century. The US had few claims to territory in banana-growing regions, but two corporations headquartered in the US set up plantations in the Latin American countries of Guatemala, Honduras and Ecuador. They were known as United Fruit and Standard Fruit then and are the predecessors of Chiquita and Dole. They controlled up to 70% of land in some countries and opposed democrat movements and post-colonial land reform initiatives and financially supported the often-repressive governments of the countries where they operated.

For this and many other reasons, including their own banana production in equally problematic colonies primarily in the Caribbean, European countries restricted imports of bananas produced by Chiquita and Dole. This persisted after their colonies became independent and turned into something called **preferential trade agreements.** They were meant to be reparations for colonization, but in practice secured a guaranteed product from former colonies and restricted economic development in other sectors. Most of the bananas were grown on small family farms and tilled by hand on hilly terrain and poor soil, with little or no mechanization or irrigation. Yields were far below those in places like Honduras, Guatemala and Ecuador. In fact, the cost of growing bananas in the Caribbean was twice that for bananas produced on the Latin American plantations. Without their preferential agreements with Europe, banana production on these small islands might have disappeared.

In the early 1990s, the single European market formed, amid deep debates about preferential and **duty-free** (or reduced taxes and tariffs) access to bananas grown in the former colonies of the Caribbean. In 1995, the **World Trade Organization** (WTO) was formed. It replaced the General Agreement on Tariffs and Trade (GATT) that, along with the International Monetary Fund and the World Bank, was a supranational institution formed after World War II to govern global commerce and trade. The WTO was designed to regulate, oversee and liberalize global trade. When European countries continued to limit and tax imports of Latin American bananas by Chiquita and Dole in favor of bananas grown in their former colonies, the US responded by imposing a stiff tariff on European goods, such as cheese and cashmere, shipped to the United States. Chiquita and Dole lodged a formal complaint with the WTO in 1997, citing that the preferential treatment of the former colonies was unfair trade.

This sparked a long and protracted conflict that ultimately, in 2012, resulted in the phasing out of tariffs on Latin American bananas over 8 years. In other words, the US went to war on behalf of one American company that already had 20% of a foreign market, and it negotiated to secure 3% of another foreign market for the benefit of seven to 10 American companies. The WTO remains a highly contentious governing body, and their meetings are protested by a wide variety of environmental and social justice activists for their rulings against workers, peasants and small-scale producers and in favor of corporations, including the disruptive Seattle protests in 1999 that halted the meetings of the ministers before they began. Caribbean banana growers now compete with global corporations for market share, which is still dominated by European companies and expats, and have turned to voluntary labeling schemes, such as organic and fair trade, to increase their profitability (Trauger, 2014).

FOCUS SECTION: JACKFRUIT

And now for a story of fruit as a form of liberation from oppression. In Sri Lanka, the jackfruit tree is called *bath gasa* or the "rice tree" in Sinhalese, shown in Figure 7.4. Rice is the primary staple grain of Sri Lanka, and prior to British colonization, Sri Lanka boasted sophisticated infrastructure for rice cultivation. In 1815, the British began occupying the island and stripped farmers of land to dedicate it to tea, spice and rubber production instead. Numerous movements to free Sri Lanka from British control ensued over the next century, but it was not until a freedom fighter, Arthur Dias, realized that without food self-sufficiency, given their inability to grow rice, the independence movements would, literally, be starved to death. Dias incorporated the planting of 1 million jackfruit trees into his independence campaigns, earning him the nickname of "Uncle Jack." The food that the jackfruit provided to the people of Sri Lanka during this time was rich in nutrients, including carbohydrates and protein and vitamins and minerals. Every part of the fruit can be eaten, including the seeds, and it can even be made into flour. The flesh takes on the texture of meat when cooked and is fast becoming a popular meat substitute in the US. It is eaten in dozens of different ways in Sri Lanka, and during the recent COVID-19 lockdown that meant loss of income and disrupted supply chains, jackfruit returned as a food staple, saving people from starvation, once again (Rathnayake, 2020).

SUMMARY

The global traffic in fruits and vegetables has its roots in colonial relationships that persist over time, resulting in inequitable access to healthy foods and political-economic relationships that perpetuate global inequalities in various ways. Food safety laws on imported food, trade agreements and technological innovations impact the production of fruits and

FIGURE 7.4 Jackfruit (https://commons.wikimedia.org/wiki/File:Jackfruit-452532.jpg)

vegetables and the hows, wheres and whys of consumption. Fruits and vegetable also have the power to transform economic or political situations due to their very nature. They are bearers of seeds for the next generation of plants, which allows for the perpetuation of the food systems, both wild and domesticated, upon which we depend.

SUPPLEMENTAL MATERIALS

Key Terms and Concepts: annual, carcinogen, dirty dozen, duty free, food scares, fruit, governance, labor, monoculture, perennial, piece rate, preferential trade agreements, profit margin, special economic zone, vegetable, World Trade Organization
Explore More: *Scientific American* published an article about produce imports to the US and created an interactive graphic to explore their origins. Explore more here: www.scientificamerican.com/article/graphic-science-where-in-the-world-your-fruits-vegetables-come-from-interactive/
Recommended Viewing: *Life and Debt* (2007), *Botany of Desire* (2009), *Fresh* (2009)
Discussion Prompt: Some countries require by law that produce be labeled with its country of origin. What is your favorite fruit or vegetable? Where does it come from? What do you know about its production? Is it certified by any agency? Aim

for 100–250 words. Be sure to correctly use at least *one concept* from this chapter *or* another chapter in your discussion post and write it in **bold** type.

Recipe: Blueberries are highly nutritious and very perishable. They freeze easily but also make a great jam. This recipe is sweetened with honey for a simple, totally local version and includes instructions for preservation: www.100daysofrealfood.com/blueberry-jam-honey-sweetened/

REFERENCES

Butterworth, M., Davis, G., Bishop, K., Reyna, L., & Rhodes, A. (2020). What is a superfood anyway? Six key ingredients for making a food "super". *Gastronomica, 20*(1), 46–58.

Freidberg, S. (2004). *French beans and food scares: Culture and commerce in an anxious age.* Oxford University Press on Demand.

Guthman, J. (2017). Life itself under contract: Rent-seeking and biopolitical devolution through partnerships in California's strawberry industry. *The Journal of Peasant Studies, 44*(1), 100–117.

McMillan, T. (2012). *The American way of eating: Undercover at Walmart, Applebee's, farm fields and the dinner table.* Simon and Schuster.

Pollan, M. (2002). *The botany of desire: A plant's-eye view of the world.* Random House Trade Paperbacks.

Rathnayake, Z. (2020). Jackfruit: The "vegan sensation" that saved Sri Lanka. *BBC.* Retrieved May 24, 2021, from www.bbc.com/travel/story/20200916-jackfruit-the-vegan-sensation-that-saved-sri-lanka.

Trauger, A. (2014). Is bigger better? The small farm imaginary and fair trade banana production in the Dominican Republic. *Annals of the Association of American Geographers, 104*(5), 1082–1100.

CHAPTER 8

Seeds

Grains, Pulses and Nuts

INTRODUCTION

In 2010, I traveled in the Indian Himalayas with climate and food activist, Vandana Shiva, to visit a remote village with a seed bank she had helped to establish. The Himalayas are home to the water supplies of the populations of both India and China, some 2 billion people. They are often referred to as the third pole, behind the north and south poles, for their rapidly changing environmental conditions due to climate change. To account for crop losses due to erratic weather that included droughts and floods in the same year, the women of this farming-dependent village saved five times as many seeds as they had in years past. The farmers shown in Figure 8.1 had begun, with Navdanya, the seed bank founded by Vandana Shiva, to save ancient varieties and integrate them into their cropping systems for greater resilience. The seed savers in the photo spoke about their experiences, with their seeds on display with a small temple to the god Shiva. Seeds are the foundation of agriculture, and as large-scale starchy crops, they form the basis of long-term food security for peasant households and national-scale wealth and power.

This chapter describes the important place of seeds, which include grains, pulses and nuts, in local and global-scale economies. Grains are staple carbohydrates, pulses are vegetarian sources of protein in the majority of the world's diets and nuts and seeds are nutritional powerhouses and alternatives to grains. This chapter discusses mechanization and economies of scale in grain production and the role of cheap food (especially in the form of carbohydrates) as a foundation for the modern state economy. Integrated cultivation and consumption practices around the world are discussed, with a focus on major commodities, some of which follow long-established patterns of exploitation and uneven development.

KEY CONCEPTS AND TERMS: GRAINS, PULSES AND NUTS

Grains are the seeds, technically fruits, of starchy crops grown and harvested for human and/or animal consumption. Commercial grain crops include **cereals**, such as corn, wheat, rice, oats, barley and rye. Whole grains include a hull, a layer that is often removed in processing to speed cooking or increase palatability or both. Grains and pulses are durable sources of carbohydrates, proteins and fats and store for long periods of time. They have

DOI: 10.4324/9781003159438-11

FIGURE 8.1 Women farmers and seed savers in Uttarakhand, India

built the foundation of agricultural economies from recorded history until today because they can be used as a kind of currency or traded as a **commodity**. They can be mechanically harvested, stored indefinitely under proper conditions, easily transported, milled for flours or refined as oils. Corn or maize, rice, soybeans and wheat are the world's most significant and valuable grain crops.

Legumes are plants that can produce a seed, known as a **pulse**, that is higher in protein than cereals. Legumes include clover, alfalfa, peas, beans, lentils, carob and peanuts, among others. They are grown for human consumption but are also used as oilseeds, green manures and for animal fodder. The pulse, or the seed that a legume produces, is unique in its symmetrical structure that opens along a seam between two sides. Legumes have a unique, symbiotic relationship with bacteria in the soil that fixes nitrogen from the atmosphere in small nodules on the plant's roots. This ability to gather nitrogen from the environment contributes to the higher protein content of its seeds, and it also contributes nitrogen to the soil, which is an important fertilizer. For this reason, soybeans are grown in crop rotation with corn on most commercial grain farms, but in other systems, such as in South Asia, legumes are grown alongside the grain crop as a green manure.

Nuts and **seeds** are a catchall category for any fruiting body with a hard shell with an edible kernel inside. Some nuts and seeds free themselves from their hard shell, and others such as pistachios must be mechanically freed. The linguistic use of the word *nut* varies from its botanical one. Seeds that are called nuts, such as almonds or pecans, are not technically nuts, but for the purposes of this discussion, nuts are seeds that differ from cereal crops and legumes in two important ways: nutritionally, they are superior to both, having more protein than a legume and more healthy fats than either, with a relatively low amount of carbohydrate, while still being energy dense. They are also often perennial crops, grown as trees, and do not require annual planting or as many chemical inputs to manage pests and disease.

FOCUS SECTION: ALMONDS

The explosive growth in alternative milks in the early 21st century fueled a boom in almond production in California. As the fastest-growing crop in the early 2000s, almonds became the most valuable commodity, exceeding wine as of 2014. Behind rice, alfalfa and corn, almonds use the most water per unit land, roughly equivalent to a gallon per nut. In a context of deepening water scarcity and increasingly severe droughts, water usage in almond production became controversial as the market exploded. Reisman (2019) queries whether almonds were a serious new threat or simply scapegoated for the problems that agriculture had also posed for California's water supplies. She argues that almonds became a real problem, and they also underscored in a very dramatic way how public resources are used for private gain, causing anxiety about power over resources and fairness in their distribution. In addition to a deep demand for public water resources, almonds rely on the honeybee for pollination. Commercial beekeepers rely on seasonal crops to feed bees year-round, and almond orchards provide needed foraging in a reciprocally agreeable arrangement between beekeepers and orchardists. Beekeepers rent out their honeybees to almond growers in February as part of a cycle of feeding that keeps bees on the move around the United States. Hazards of transportation, stress and exposure to parasites and chemicals in the environment through these practices threaten honeybee health and can contribute to colony declines (Traynor, 2017).

Nuts and seeds are sometimes considered a grain and are quickly becoming a healthier, potentially more environmentally sound alternative to annual grain crops with superior nutritional properties. And for many with intolerances to grains, nuts are a key replacement, especially as alternative milk. According to Martin Crawford, a noted permaculturalist in the UK, sweet chestnut trees can yield up to 2 tons per acre, which is the equivalent of an average yield for organic wheat. Sweet chestnuts have a nearly identical caloric profile to grains, with greater nutritional value and health benefits and higher yield per acre with far fewer inputs. According to Molnar et al. (2013) a small city with a population of 47,000 could obtain about 200 calories per person per day from walnuts grown on only 2.5 acres. Figure 8.2 shows chestnuts for sale in a market in Slovenia.

Most of the world meets its dietary needs for carbohydrates through annual crops of wheat, rice and corn. The Green Revolution was instrumental in transforming the

FIGURE 8.2 Chestnuts at Ljubljana Central Market, Slovenia (https://commons.wikimedia.org/wiki/File:Chestnuts_at_Ljubljana_Central_Market.JPG
Conceptual Engagement: Commercial Cereal and Legume Production)

landraces most of the world grew and standardizing them through selective breeding. The so-called improved varieties were bred to have higher productivity, ripen at the same time and be appropriate for mechanization. The increasing capital demands for irrigation, chemical inputs and tractors led to an increase in the size of farms and land consolidation that is an ongoing feature of agricultural industrialization everywhere. In addition to technological change that allowed for expansion, in the 1970s states pushed farmers to expand production. This process is called an **economy of scale**, wherein large quantities of a single thing can be generated on a large piece of land through standardization and mechanization. Figure 8.3 shows a combine harvesting soybeans, demonstrating the scale of machinery required for modern grain harvesting. Grains and pulses are important as food for humans, although corn and soybeans are widely used as animal feed and for industrial products, including inks and lubricants. Fuels such as ethanol, made from corn, are increasingly becoming central pillars of domestic production as countries seek to develop alternative fuels. The overproduction of some grains is key to foreign policy as an instrument of food aid and through trade agreements.

The importance of grains and legumes to an industrial and industrialized agricultural economy and foreign policy means that commodities such as grains and pulses are supported by **subsidies**, or direct payments of federal money to farmers to grow certain crops. Many rich countries make direct payments to farmers as a way to support the agricultural industry because the margins on food production are quite low. The margins on the crops are so low (sometimes not breaking even) that farmers would grow other crops without such incentives. This is part of something called a **cheap food policy**, which although controversial, is intended to lower the cost of food (grains and meat and other industrial products) so that wages in the industrial sector may be kept low. This allows industry to accumulate profit by suppressing wages below what labor is actually

FIGURE 8.3 Machinery used to harvest soybeans

worth, but people can still eat. Remember how concerned Borlaug was about revolution following famine? Well, the same logic applies here: hungry people will protest their political-economic conditions sooner or later, and people cannot eat if they cannot buy food.

Because of their widespread production and standardized genetics after the Green Revolution, corn, soybeans, rice and wheat became increasingly vulnerable to pests. The importance of these crops to other industries also required a level of uniformity of product not often found in nature, such as the absence of weed seeds, the presence of which results in significant reductions in the price. In the 1980s, biotechnology offered solutions to both problems. Corporations such as Monsanto worked with government agencies and universities to develop genetically modified (GM) grains and pulses that could resist pests, disease and herbicides. The technology is relatively simple: a bacteria or virus is grown in a lab with resistance to an agricultural chemical, for example, and those genes that render resistance are spliced into the DNA of the target, now patented crop. The use of such seeds revolutionized agriculture and began an uncontrolled environmental experiment that some countries in Europe and Africa are unwilling to participate in, no matter what the cost in trade and aid. Some farmers also resist the use of the seeds, and even when their crops are contaminated, companies sue for patent violation and win, such as in the case of Percy Schmeiser in Canada. Many people fear the health risks associated with genetic modification technology, but research has shown them to be relatively low. The costs to autonomy for entire nations, individual farmers and the genetic future of seeds is of more concern to some experts, however.

Agricultural biotechnology remains a controversial method of technology transfer to developing countries, and many have refused GM crops either as aid or as investments. Kenya is one such country that initially trialed GM corn and then put in place a ban to protect its trade relationships with the European Union, which bans the import of GM crops. There is considerable debate over the effectiveness of agricultural biotechnology to create food security in developing countries (Harsh & Smith, 2007). Consistent with past interventions in poverty and hunger, richer, wealthier countries have led the growth of biotechnology sectors, often as a conditionality for other forms of aid. In the early 1990s, Monsanto, the Rockefeller Foundation, the World Bank and USAID launched a series of biotechnology projects in Kenya, a country that at the time lacked strong government oversight over the introduction of such varieties. Like the Green Revolution, genetically engineered seeds are not necessarily meant to address issues of food security, although they are often pitched that way to international donors. Like the Green Revolution used the specter of famine to enroll farmers in circuits of global capital, food security is often stated as the goal behind introducing biotechnology to places where it was not yet established. The market share for Monsanto's GM corn in the Americas was dwindling, and in a bid to capture more markets, Monsanto introduced its crops to African countries in the guise of philanthropy. This is arguably a strategy of Harvey's (1990) **spatial fix** (Parker, 2017).

EXAMPLES

Wheat-Rice in India

The wheat-rice cropping system is practiced in South Asia in Bangladesh, India, Nepal and Pakistan in what is known as the Indo-Gangetic Plains (IGP) and into the lower to middle hills of the Himalayas (Bhatia, 1965). Historically, wheat and rice were grown in different regions of the IGP due to their long growing seasons, but the introduction of improved varieties through the Green Revolution made it possible to shorten the growing seasons of both and plant them in rotation throughout the year. Rice is grown in the rainy season of the monsoon climate, which is most of the summer, so that the rice can be rain fed. Wheat is grown in the winter during the dry season. This cropping system is now the dominant agricultural system throughout the IGP on terraced hillsides such as the one seen in Figure 8.4. The farmers and seed savers who opened this chapter grow soybeans on the edges of their terraces as a form of nitrogen fixing green manure, which supplements and compliments the grains in their diets. They are also a visual representation of the **feminization of agriculture** in much of the world. While the labor of agriculture has always been performed by women, men are increasingly migrating to cities in search of waged work, leaving women to perform all of the work of agriculture.

FIGURE 8.4 Rice and soybean crop terracing in Uttarakhand, India

FOCUS SECTION: PINE NUTS

Ever wonder why pine nuts are so expensive? Pound for pound, they are one of the more expensive food items in the average grocery store, at about US$20 per pound, more than three times the cost of other relatively expensive nuts, such as almonds and cashews. Slightly sweet and buttery, they are a staple of the Italian sauce, pesto, which is made with basil, parmesan and olive oil. Pine nuts are the edible seeds of a few species of pine trees and are extracted by hand from the protective shell of a pine cone. Four species, which are primarily uncultivated in forests, compose the bulk of global pine nut production: Colorado pinon, Mexican pinon, Chinese nut pine and Italian stone pine. Trees take up to 25 years for mature production and more than 18 months to ripen. The hard outer shell and an inner shell must be removed by hand during harvesting and processing. Unsurprisingly, pine nut production is commonly undertaken in countries with low labor costs, such as Pakistan and China, and some with illegal labor practices, such as North Korea. Unlike fruits and vegetables, however, the country-of-origin label is not required for seeds, so it is difficult to know the provenance of contemporary pine nuts.

Millets in West Africa

It is said in west Africa that you haven't eaten a meal if you haven't eaten millet with it. Millet is a staple crop of several west African nations in the semi-arid Sahel region, just south of the Sahara Desert. Africa is a key center of domestication for a wide variety of cereal crops, including two millets, pearl and finger. Millets comprise a broad category of plant seeds that are composed of hundreds of varieties that retain genetic diversity that is adapted to a wide range of environmental conditions. The superior nutritional quality of millets, which includes carbohydrates, protein and vital micronutrients, plays a critical role in the food security of the region, the cultivation of which relies on the retention of historical practices of cultivation, few inputs and customary ecological knowledge. Climate change threatens the already fragile but fertile soils of the Sahel and the communities who directly rely on agriculture to meet their nutritional and livelihood needs. Individual communities retain varieties unique to their villages and thus hold key genetic information that will be essential to cope with changing environmental conditions. Like the seed keepers in India, women are the holders of genetic and cultivation information and are now increasingly responsible for managing small-scale farms in the ongoing feminization of agriculture taking place here as well (Kerr, 2014).

Quinoa

Like the millets, quinoa, which is technically a pseudo-cereal, is a small seed that is a nutritional powerhouse. Quinoa and its cousin, amaranth, are pseudo-cereals, not grains, cultivated in extreme conditions in the Andes mountains of South America. Wide ranges of temperature, low oxygen levels and poor soils are conditions in which it thrives, and little else does. Packed with protein, carbohydrates and micronutrients, quinoa is traditionally a sacred food crop to the Indigenous people of the Andes. Since the 1990s, global demand for quinoa has exploded, with major markets in the European Union, the US and parts of east Asia. While the growth of foreign markets, and a near tripling of its price, has boosted the incomes of growers in Bolivia and Peru, it has become scarcer on the tables of those who grow it because they can no longer afford to eat it. As the theory of uneven development would tell us, development in one place comes at a price in another, which has the effect of developing a two-class food system in which the global consumer enjoys a diet nutritionally superior to those locals who produce their food. In addition to an abrupt and dramatic change to diet that has historically been devastating to Indigenous people, land disputes have erupted as privatization and consolidation drive smaller-scale farmers off the land, and larger-scale producers capitalize on the new global market (Bedoya-Perales et al., 2018). As agriculture modernizes, people, usually men, migrate to urban areas, putting them in a wage-labor relationship to buying food, which in many cases, they can no longer afford. This contributes to food insecurity in a cradle of domestication and characterizes one of the major contradictions of the food system.

Soybeans in Argentina

Nearly half of Argentina's arable land is planted in soybeans, almost all of it transgenic or genetically modified to resist herbicide applications. The soybeans were planted for export largely to markets for livestock feed in Southeast Asia and have become Argentina's largest export by value. The rapid expansion of agribusiness in Argentina since the 1990s, due to the liberalization of trade, brought the eviction of peasant and Indigenous farmers from their lands, environmental devastation and the predictable concentration of financial and political power in the hands of already wealthy landowners and companies. The expansion of soybean production in the Pampas, a region of Argentina dedicated to export-oriented commercial agriculture, involved massive applications of pesticides to large swaths of the landscape. Some of it, documented by Lapegna (2016), drifted into neighboring peasant communities that were negatively impacted in terms of their own health and their crops. The response was widespread protests that were effectively quashed by the government at the behest of foreign multinationals and included the killing of Indigenous activists. GM crops are often hailed as the answer to hunger and poverty, but the example of soy in Argentina illustrates that like the Green Revolution, the introduction of technology and capital to agricultural problems further exacerbates the existing problems and creates new ones.

Wild Rice

Wild rice (*Zizania palustris*) is a grain endemic to the lakes and ponds of the Upper Midwest of the US and parts of southern and eastern Canada and has provided a staple food

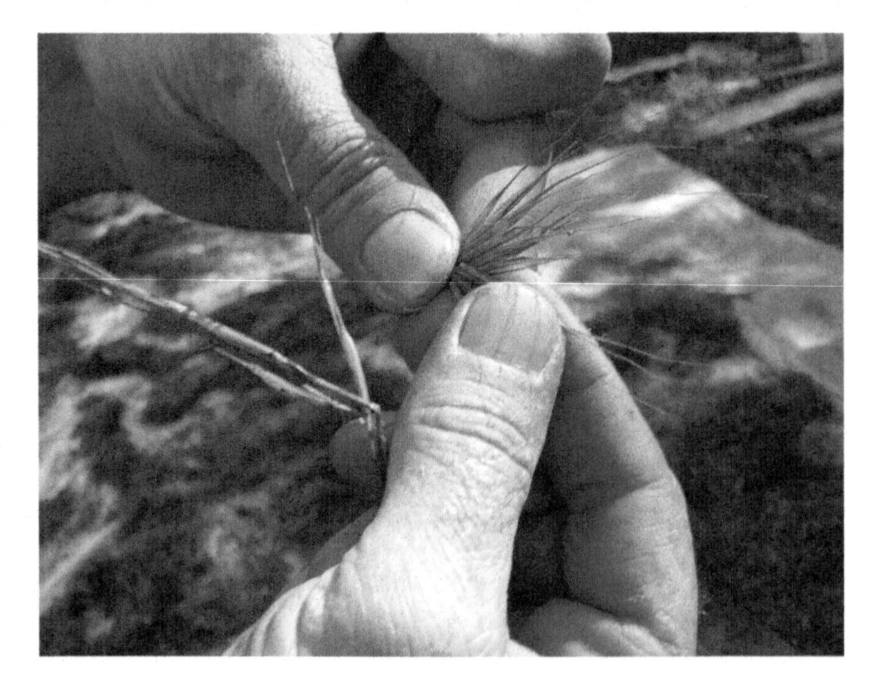

FIGURE 8.5 Wild rice grains

for Anishinabek and other tribes living there for centuries. It is a species of aquatic grass, unrelated to modern cultivated white rice (*Oryza sativa*). *Manoomin*, as it is called by the Anishinabek, shown in Figure 8.5, has not been bred for commercial production until recently, and the rice in northern lakes and waterbodies has evolved over time to be specific to watersheds, producing at least two distinct subspecies. Wild rice is sensitive to environmental disturbances, including and especially pollutants. As a wild perennial, *manoomin* does not require cultivation, but native ricers often deliberately drop grain into the water as they harvest in a spiritual-ecological practice referred to as "seeding" to ensure the next year's harvest, also known as pre-domestication cultivation. Wild rice is a high-fiber, high-protein, low-glycemic whole grain that like quinoa is a nutritional superfood. It is harvested in canoes in the autumn and then parched or dried in wood-fired kilns. The wild rice that is available in the grocery store is a domesticated variety that is patented and which many tribes have fought (LaDuke, 1994).

FOCUS SECTION: BEER AND BREAD

In Germany, different types of beer were developed in different regions according to the availability of the raw materials for making beer, local conditions and customs, and methods of fermentation (top or bottom, which gives us ales and lagers, the two broad subtypes of beer, respectively). The Kolsch is brewed in Koln, the Pilsner (today's Budweiser, Michelob, etc.) in Pilsen and the Schwarzbier (or black beer) in Bavaria. Distinctive styles of bread were developed alongside different styles of beer. Germany boasts some 300 different breads and over 1,200 different forms of baked goods. One of the more famous of German breads, Westphalian pumpernickel, is made of whole grain rye and steamed in long tubes for 24 hours. This geography of bread relates to the way spent grains (wheat, barley and rye) from the brewing process can be used to bake breads, but also because of the Bavarian order of 1516, called *Reinheitsgebot* or the German Purity law. The law restricts beer manufacture to barley, hops and water (yeast was added later when its role became clear). One of the first food safety laws on record, the Purity Law has remained largely unchanged, and while it influences other nation's beer brewing practices, it does not apply to beer brewed outside of Germany. The Purity Law was introduced in part to protect the baking supply chain, which relied primarily on wheat and rye, and to keep those grains from being used to make beer. This also kept the price of staple foods free from competition from breweries and therefore low, which as we've seen many times over, is key to ensuring a compliant labor force.

SUMMARY

Seeds, in the form of grains, legume and nuts, form the basis of household- and global-scale food security. They are diverse, highly adaptable and widely consumed in a variety of products. Their production is subject to the highest level of political-economic interventions and technological innovation to produce as much as possible with the most efficiency. Seeds are not only food for people, they are fuel, animal feed and the basis for a growing number of industrial products. Their production, however, is costly in industrial contexts and relies on subsidies from wealthy and powerful states. In small-scale peasant economies, the export of essential crops to feed people in the Global North is a driver of food insecurity. Women in vastly different places are increasingly responsible for growing grains for household consumption as men migrate to cities for wage work in an ongoing modernization of food-producing economies everywhere.

SUPPLEMENTAL MATERIALS

Key Terms and Concepts: cereals, cheap food policy, commodity, economy of scale, feminization of agriculture, grains, legumes, nuts, pulses, seeds, spatial fix, subsidy

Explore More: The Foreign Agricultural Service of the US Department of Agriculture creates maps of crop production globally. Explore the Crop Production Maps under the Crop Calendar and Production Maps tab here: https://ipad.fas.usda.gov/

Recommended Viewing: *King Corn* (2007)

Discussion Prompt: The Maya are thought to be "People of the Corn" because it defined their lives in so many different ways. Examine your diet for the grains that you consume. What nation of grain are you a person of? Why do you think this is? Aim for 100–250 words. Be sure to correctly use at least *one concept* from this chapter *or* another chapter in your discussion post and write it in **bold** type.

Recipe: I love pesto, but I don't love the expensive imported ingredients a traditional Italian pesto calls for. By experimenting a bit on my own one summer, I came up with a delicious and dairy-free version using sweet cherry tomatoes and pecans. You can find a similar recipe here: www.gabriellemyers.com/gluten-and-dairy-free-recipe-blog/sungold-basil-and-pecan-pesto

REFERENCES

Bedoya-Perales, N. S., Pumi, G., Talamini, E., & Padula, A. D. (2018). The quinoa boom in Peru: Will land competition threaten sustainability in one of the cradles of agriculture? *Land Use Policy, 79*, 475–480.

Bhatia, S. S. (1965). Patterns of crop concentration and diversification in India. *Economic Geography, 41*(1), 39–56.

Harsh, M., & Smith, J. (2007). Technology, governance and place: Situating biotechnology in Kenya. *Science and Public Policy, 34*(4), 251–260.

Harvey, D. (1990). *The condition of postmodernity: An enquiry into the conditions of cultural change*. Blackwell.

Kerr, R. B. (2014). Lost and found crops: Agrobiodiversity, Indigenous knowledge, and a feminist political ecology of sorghum and finger millet in northern Malawi. *Annals of the Association of American Geographers, 104*(3), 577–593.

LaDuke, W. (1994). Traditional ecological knowledge and environmental futures. *Colorado Journal of International Environmental Law and Policy*, *5*, 127.

Lapegna, P. (2016). *Soybeans and power: Genetically modified crops, environmental politics, and social movements in Argentina*. Oxford University Press.

Molnar, T. J., Kahn, P. C., Ford, T. M., Funk, C. J., & Funk, C. R. (2013). Tree crops, a permanent agriculture: Concepts from the past for a sustainable future. *Resources*, *2*(4), 457–488.

Parker, L. D. (2017). *Modified states: Sovereignty and the ethics of crop biotechnology in Kenya* [Doctoral dissertation, Department of Geography, University of Georgia].

Reisman, E. (2019). The great almond debate: A subtle double movement in California water. *Geoforum*, *104*, 137–146.

Traynor, J. (2017). A history of almond pollination in California. *Bee World*, *94*(3), 69–79.

CHAPTER 9

Protein

Meat, Fish, Dairy and Eggs

INTRODUCTION

The European Society for Rural Sociology holds annual conferences in different parts of Europe each year. During the middle of the conference week, on Wednesday, conference participants can choose from a variety of food- and agriculture-themed tours for the day. In 2004 when the conference was in Norway, I chose a tour of a sustainable mountain dairy farm, with some very pampered cows, that culminated in a feast of traditional Norwegian foods. Long before charcuterie boards became all the rage in the US, Nordic and Germanic cultures had meats and cheese served on a board as a way of life for centuries. No stranger to the weirder Norwegian foods of my childhood, I wasn't surprised by any of the pickled fish, boiled beets, thick potato dumplings or sweet, brown cheeses on a long table loaded with food, but what did shock me was a large plate of reindeer tongue. Traditionally herded by the Indigenous peoples of the Nordic countries in semi-wild herds, as seen in Figure 9.1, reindeer is a sustainable and healthy meat, and obviously, no part of the deer is wasted in the consuming of it.

Modern systems of meat and protein production could not be more different from this, as is explained in what follows. This chapter focuses on the production of animal proteins throughout the world with sections on meat, dairy products, fish and eggs. This chapter discusses the sustainability and ethics of different production practices, technological innovations, and governance of systems associated with animal protein consumption. The chapter also addresses protectionist trade policies, aquaculture trends and the biopolitics of pasteurization and local markets.

KEY CONCEPTS AND TERMS: ANIMAL PROTEINS

Meat, Poultry and Fish

Meat and **fish** were vital complements to foraged grains, nuts and berries in human diets for millennia. Some evidence also suggests that animal flesh assumed some ceremonial and political significance in early human societies, which persists in some cultures today. Much like the way domestication irrevocably influenced the social structures of human cultures, the use of fire to cook food, especially meat, changed

DOI: 10.4324/9781003159438-12

FIGURE 9.1 Reindeer herding in Sweden (https://commons.wikimedia.org/wiki/File:Reindeer_herding.jpg)

humans physiologically. Energy resources previously used to fuel digestion in a long and complicated gut, were diverted to the growth of a bigger brain instead when cooked (pre-digested) food was consumed (Wrangham, 2009). While animal domestication lagged behind plants in most places, the selective culling of some animals eventually allowed the more docile and palatable species to be brought closer to human populations and their settlements.

Today, humans consume a wide variety of red meat, the vast majority of it in agricultural production, including beef, lamb, pork, water buffalo and goat. Wild animals in the form of venison, reindeer and bushmeat are also widely consumed. Humans also eat **poultry** or other birds in the form of turkey and chicken, emu, quail and duck. We also consume a vast number of freshwater and saltwater aquatic species, caught wild, or raised through **aquaculture**, or fish farming. In general, animal consumption increases with wealth, and many of the world's poor eat animals only rarely. Figure 9.2 maps daily meat consumption throughout the world, and it follows a strong pattern of economic development and power within the supply chain. Expected meat consumption according to European Union standards is between 40 and 165 g/day, with wealthier countries exceeding that standard and poorer countries not reaching it. In wealthy countries, the widespread consumption of meat as a commodity is supported by enormous subsidies of grain production and fossil fuels, both of which are necessary to produce large quantities of inexpensive animal products on a large scale.

Daily meat consumption per person, 2013

Average daily meat consumption per person, measured in grams per person per day. Countries with daily meat consumption greater than the expected EU average of 165g per person are shown in red; yellow are those countries below 165g but exceeding the more ambitious limit of 40g per person; and in blue are those below 40g per person.

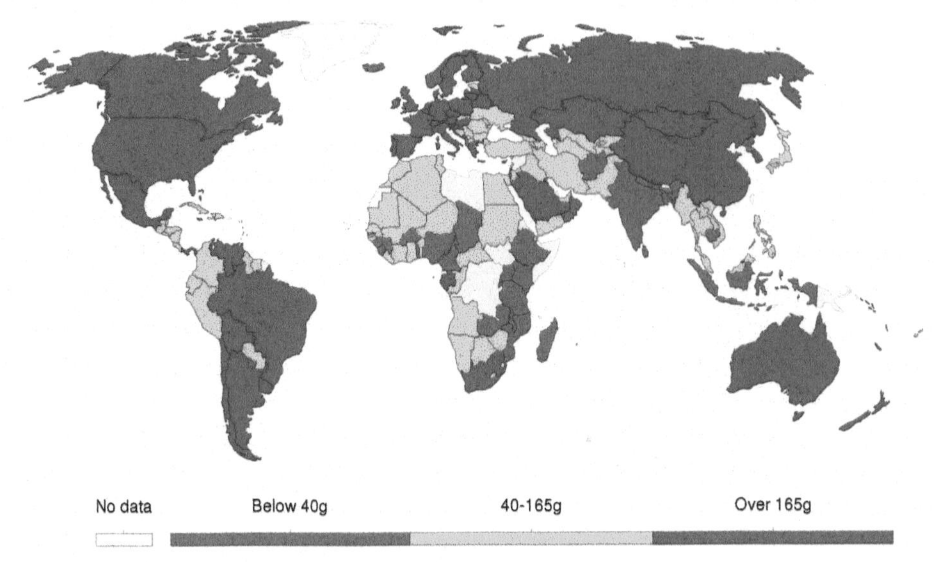

Source: UN Food and Agricultural Organization (FAO)

FIGURE 9.2 Daily meat consumption per person (https://en.wikipedia.org/wiki/List_of_countries_by_meat_consumption)

Milk and Milk Drinking

The ability to digest **lactose**, the sugar in milk, is thought to have evolved in human populations around 7,500 years ago in Central Europe. It is believed to be an adaptation to cattle husbandry and a lack of access to other sources of vitamin D, protein and carbohydrates in the lean months of late winter. Livestock could be bred in the fall and summer and consumed stored forages throughout the winter to give birth in early spring when other sources of food were exhausted. The evolution of milk drinking required a metabolic mutation referred to as **lactase persistence** that allowed humans to drink milk after they matured to adulthood. Lactase persistence is not universal in human populations who consume milk as food, however. Most Asians, Africans and Indigenous people of America and Australia are lactose intolerant, but Central Asians rely on fermented milk, which has a lower level of lactose than milk. Cheese and other fermented dairy products such as yogurt and kefir are also relatively low in lactose, can be transported by a pastoral society and are less perishable. It is likely that milk drinking favored settled populations more than pastoral ones, thus explaining at least some of the variations in lactose intolerance globally. Alternative milks made from nuts or grains, for allergy, intolerance or ethical reasons, have risen in manufacture in recent years, but they are nutritionally inferior and far less amenable to processing into other dairy products than goat, sheep and cow milks.

FOCUS SECTION: VEGANISM

The consumption of animal proteins leads to concerns about the ethics and morality of such practices. The environmental consequences that include contributions to climate change and animal cruelty in some production models, and the ethics of eating animals at all, fuel intense debate about whether and how to eat animal protein. While there is some consensus about the human health and environmental benefits of reducing red meat consumption in particular, there is no consensus on how to replace the necessary macronutrient protein in human diets. While plant-based protein substitutes have never been more available and tasty, they are expensive, often highly processed and have long supply chains associated with them. Lab-grown meats, discussed more later in this chapter, are an ethical alternative that Mouat et al. (2019) argue turns moral concerns into capital with an energy-intensive but highly profitable new product. Harper (2011) argues that all oppressions are linked, and that veganism is a path toward societal healing when its politics takes on anti-racist consciousness. Indigenous scholars have also critiqued veganism for its whiteness as well the way plant-based foods are inaccessible to many, particularly the poor and people of color. But the associations with agriculture and colonialism and a long history of sustainable harvest of wildlife inform many Indigenous people to seek alternatives to industrial animal proteins (Dunn, 2019). In short, there is no completely ethical approach to consuming any food, because everything has a trade-off. With animal proteins, the stakes are higher, but the simplest answer may lie in diversifying protein sources, reducing the amount of animal protein and seeking out as many protein sources in local supply chains as possible.

Eggs

Humans have also eaten eggs of many species of birds and fish for millennia. Eggs are nutritional powerhouses packaged in a hard shell, like seeds, and are an inexpensive source of protein, fat and essential micronutrients, such as amino acids and vitamins. Easily foraged from wild birds, fish and reptiles in the human past, they have persisted into the present as a staple food. Now, humans eat primarily chicken eggs, which are an inexpensive and healthy source of protein and fats. Chickens can lay up to six eggs each week, which makes them more productive than most protein sources. Eggs are encased in an impermeable barrier called **albumin**, and as long as they are not washed or fertilized, they are stable at room temperature for many weeks. Only in a few countries are eggs treated as perishable, requiring a **cold chain** (refrigeration from production to consumption), which also makes them a low-carbon protein in most places. Hardy and adaptable, chickens can eat almost anything and are raised in a staggering array of conditions around the world: from free-ranging birds in the small towns and megacities of the developing world to massive barns filled with thousands of caged birds in the developed world. In an appeal to consumers, eggs are now marketed under a wide range of labels attempting to signal ethical animal treatment and healthier food products. These labels include cage free, free range, organic, Omega-6 and so on, that do little to reveal the conditions of production but are used to comfort and lull

FIGURE 9.3 Eggs from the author's backyard chickens

consumers. Different breeds of chickens lay different colors of eggs, which explains the variety you see in Figure 9.3 from my chickens. These are from a mix of free-ranging Marans, Auracana, Speckled Sussex, Wyandotte and Cream Legbar chickens.

CONCEPTUAL ENGAGEMENT: LAND USE PRACTICES AND RATIONALIZATION

The US, Australia, Brazil and Argentina dedicate large tracts of land to the production of livestock. Some of this is explained by the land required to graze cattle in semi-arid grasslands, but another significant source of land use is devoted to feedlots and producing food crops for confined animals. A relatively small amount of land in these countries is dedicated to the growing of food we actually eat, such as fruits, vegetable, nuts and cereals. According to an analysis conducted by Peters et al. (2016), the carrying capacity of different diet scenarios suggests that both meat-eating and strict non-meat-eating diets are both unsustainable in terms of land use, and they identify lacto-omnivore diets (diets that include milk and meat in addition to plant-based protein) as the most sustainable. Dairy cattle require significant inputs of land and resources, but one cow produces a lot of milk, which can be turned into a wide variety of products with long shelf lives, and dairy cows can be raised sustainably on pasture. The reason why this diet requires an omnivorous pattern is because you do not get milk without a calf, half of whom are males. Most male calves are useless in a dairy system and are used for veal production. The lacto-omnivore diet does not include other kinds of red meat, or cattle raised solely for beef, which experts agree is detrimental to the environment in a multitude of ways when raised in industrial systems.

The majority of the world's agricultural systems follow such a diet with few exceptions. Many people in rural India and Africa rely on eggs and dairy products for their major protein sources and only eat animal flesh on special occasions. Domesticated animals in these contexts are often raised in an **integrated crop-livestock system** within which animals and plants are raised together as complementary forms of production. The center of an integrated system is the herd of grazing animals, ruminants such as cattle, sheep or goats, which graze on grass or other crops and generate nutrients in manure, which is then used to fertilize the soil for crops. Researchers in Kenya demonstrated how this system can sequester carbon in a system to mitigate climate change and also move nitrogen through a closed loop so that it does not contaminate water supplies (Ortiz-Gonzalo et al., 2017). This system supplies the vital macro- and micronutrients to consumers of its products as well, and in some places, the animals provide the traction power for cultivation, negating a need for fossil fuels to be added to the system. This complex system also requires access to and control over land, which has been disrupted by colonialism, privatization and state-based **territorialization** (the practice of claiming and using space) of land for national-scale uses.

Modern, industrialized systems of meat production could not be further from this system. While still connected in some ways through **nutrient management**, or the use of manure on crop fields, animal production takes place in an isolated indoor environment within which animals are fed crops, often mixed with supplements, antibiotics and hormones to mitigate the adverse impacts of confinement. The development of confined animal feedlot operations, or **CAFOs**, is a form of Max Weber's theory of **rationalization** – the establishment of quantifiable sets of rules to improve efficiency and productivity which rejects decision-making on the basis of emotion, wisdom, tradition, Indigenous knowledge and the like. It is a hallmark of modernity, Western society and capitalist markets. George Ritzer (2013) referred to this tendency as **McDonaldization** of society and identified four characteristics: efficiency, calculability, predictability and control.

Practices adhering to these characteristics operate to maximize the amount of meat or milk an animal produces while keeping the levels of inputs (feed, supplements) at their optimal (and most efficient) level so that the product will be the same result every time. McDonaldization and rationalization are not seen as problematic to most people; rather they see a uniform, predictable product to be comforting, desirable and reassuring. A fast-food burger will be the same wherever you buy it. That is, until it isn't. The rationalization of protein in the 1980s in the UK and the US led to the consumption of rendered cattle by cattle, leading to an outbreak of **bovine spongiform encephalitis** that can lead to a debilitating, degenerative, fatal brain disease in humans and required the slaughter of thousands of cows. Natural systems resist uniformity and rationalization, and the human consequences of forcing them into it can be disastrous.

Presented in what follows are four examples of responses to the social and environmental challenges that are created through the production of animal proteins for human consumption. The first is labeling schemes that seek to ground responsibility and accountability through state-based governance of food production. The second is fish farming as a solution to overfishing and health problems associated with meat consumption, with unintended consequences for human and environmental health. The third example regards the consumption of milk and the power relationships that shape the supply chain for milk protein in wealthy countries. The last example is local markets for meat and other animal proteins which seek to change the power relationships between producers and consumers by developing short supply chains.

EXAMPLES

Labeling Schemes

An effort to combat the degradation of quality from rationalization is **geographic indications**, which provide information about the origin and quality of food to consumers. One such label is the **protected designation of origin** (PDO) in Europe that requires the entire production supply chain (including processing) for a given food item to take place within a designated region. The system includes extensive auditing, and auditing and is protected by law. This is a strategy to guarantee the quality of a given product when it is associated with a geographic origin. For example, Parma ham, a raw, aged meat similar to prosciutto, has been produced in a particular way in Parma, Italy, for centuries. The artisanal producers want to retain the quality of the product and not have it undermined by another, inferior product that could bear the same name. A supply chain analysis conducted on geographic indications found that such labeling schemes were able to add value to products at the retail level, but that value accruing to producers was mixed and dependent on a variety of factors (Cei et al., 2018).

Labeling laws elsewhere, such as **Country of Origin Labeling** (COOL), require retailers to inform their customers about the source of certain foods. It is applied in the US to fruits and vegetables and most recently to meat and meat products. It is a **protectionist** trade policy, or one that is designed to protect the markets in the country where the product is sold. The thinking is that citizens will prefer, if given the knowledge and the choice, to purchase food items from their own country. It was also put in place to protect American farmers from low-cost imports of meat from countries such as China, who have very different labor and environmental laws governing production. COOL is a controversial topic, with advocates and opponents on both sides of the political spectrum. It is often a political hot potato and a negotiating chip in terms of trade with foreign countries.

Fish as Health Food?

Since the 1990s, fish has become the healthier protein, since it is lower in saturated fats and higher in health fats than every other animal-based protein. Since then, the wild fish catch of the world has stagnated and declined relative to the growth of aquaculture. Practiced for millennia in Asia and elsewhere, **aquaculture** is the practice of raising fish in controlled conditions. Fish have long been integrated into human cropping systems, such as the introduction of carp to rice fields in China. The recent intensification of fish production, the so-called **blue revolution**, has taken its cue from the rationalization of agriculture and intensified the conditions under which fish are raised in captivity, and along with it, pollution, disease and food scares. Ninety percent of farmed fish are raised in Asia, with China the leader in production and export. Tilapia and salmon are two species commonly found in fish farms and sold in markets around the world. Fed a diet of mostly grains and soybeans, fish grow quickly and require a fraction of the food that livestock need to produce the same amount of protein.

The problems of disease and pollution, as well as quality, give some consumers pause. The sustainability of wild caught fish is also questionable due to overfishing, habitat loss, climate change and pollution. Non-profits try to educate consumers about the best fish to eat and also certify some fisheries as sustainable. Generally, sardines, anchovies, cod, oysters, catfish, tilapia and some farmed salmon are good choices. Some wild and farmed fish are accompanied by consumption advisories for pregnant women due to the heavy load of heavy metals in their flesh, calling into question both the human health aspects of fish consumption as well as the environmental health of certain production practices. A sustainable alternative was developed at Growing Power in Milwaukee, Wisconsin. As part of an integrated system of fish and crop production in greenhouses, known as **aquaponics**, tilapia are raised in tanks below greens, which grow in the nutrient-rich wastewater from the fish, also cleaning the tanks as they grow. The fish are sold along with the sprouts and greens in local markets in a food-insecure neighborhood. The Community Seafood Movement is also gaining steam, with small-scale fisheries offering fresh-caught seafood on a subscription basis to individual consumers. This direct marketing scheme skips the long, complicated and expensive supply chain and delivers fish fresh from the ocean to the consumer's doorstep. Still in its infancy, consumer demand could drive the resurgence of fishing as a livelihood.

FOCUS SECTION: CHILE AND FISH FARMING

Off the coast of southern Chile, in the scenic fjord lands region of Patagonia, lies Chiloe Island, parts of which are in a protected bioreserve. The cultural hearth of potato genetics, Chilean peasant farmers have grown potatoes using the kelp from the nearby ocean as fertilizer. In the 1990s, salmon aquaculture came via international fishery corporations to Chile and became the fastest-growing industry in the country. Chile became the second largest producer of farmed salmon in the world, the majority of which comes from Chiloe. The protected bays between the island and the mainland made an ideal fish farming environment, and the fish farms attracted many potato farmers into waged work in the salmon pens. The overcrowded conditions in the pens and the lack of regulation by the Chilean state led to outbreaks of viruses and mass die-offs in 2016. The dead fish accumulated in the pristine waters of the bay and contaminated the entire region, contributing to die-offs of other species. The collapse of the industry led to the layoffs of workers, and the contamination of traditional fishing grounds plunged the local communities into poverty and food insecurity. Mass protests resulted in demands on the Chilean state for assistance. The state did little in response, but the industry has since rebounded somewhat, with little in the way of regulatory change, potentially setting the stage for further disasters (Gerhart, 2017).

Raw Milk and Production Networks

Pasteurization is a heating process that slows the microbial growth in foods. Invented by Louis Pasteur in1864 to extend the shelf life of alcoholic beverages, it ensures some level of food safety – mostly from industrial production practices that introduce and promote the growth of pathogens. Industrial era milk was thought be unsafe from about the 1850s onward, due to bacteria found in cattle living in unsanitary conditions, which is easily transmitted to humans. One of the earliest food scares was related to the fact that infant mortality increased when babies were fed contaminated cow milk. Without product testing technology, which was not available before 1910, pasteurization became an important tool in reducing infant mortality. Pasteurization persisted even when improvements to production systems made milk safer, testing was widely available, vaccinations made animals healthier and food safety laws made milk cleaner.

Pasteurization equipment is expensive and complicated and is located at the processing, bottling and distribution point in the supply chain, giving processors the power to determine the supply chain. This control is supported by state-based food safety laws that govern the presence or absence of pathogens and microbes in milk. Raw milk advocates argue that milk is as safe as the dairy it came from and that beneficial bacteria – those used to make yogurt – outcompete pathogens when they are allowed to be present. Advocates therefore argue that raw milk is healthier, maybe safer, and that the laws that forbid raw milk sales are designed to benefit the companies that control the equipment and the distribution. Most milk consumption in the world is of unpasteurized products, such as the milk in the water buffalo I'm milking in Figure 9.4, which we used in our tea a few minutes later.

Local Markets and Meat

Recall from Chapter 6 that the advocates for food sovereignty, like Joel Salatin (2007), a farmer, writer and activist, argue for local food production, food cooperatives, local

FIGURE 9.4 The author milking a water buffalo in India

processing and solidarity economies as key mechanisms toward autonomy in the food system. Salatin, a polarizing figure within the food sovereignty movement, is a forceful advocate for freedom for farmers to grow, process and sell food to consumers without government oversight, launching a libertarian argument that consumers should be allowed to engage in risk if they want to. Recent consolidation in the meat packing industry made plants certified by the US Department of Agriculture difficult to access for most small-scale farmers. In addition, government regulations restrict the on-farm processing of animals destined for markets. Salatin, like many farmers, used to process his meat chickens on his own farm and without access to affordable processing options feels forced out of business. Salatin argues that people need to be allowed to take their own risks and eat food that farmers produce, however it is processed.

Not all food sovereigntists agree with this approach, but they do advocate for and desire "policies that protect local production and markets" and that work to "eliminate corporate control and facilitate community control" (Nyéléni, 2007, p. 27). This could be accomplished, according to scholars and activists, via the conscious and deliberate (re) negotiation of foundational economic ideas and practices, the development of new economic languages, and the creation of new economic subjects. This strategy also engages with the principle of **subsidiarity**, which encourages decision-making at the lowest scale and between the fewest people possible, such as agreements between producers and consumers about what food is safe to consume (Feagan, 2007). For many food sovereigntists, food safety can come about through the market, state and civil society acting to increase access to a wide variety of safe foods, not by restricting access to food in ways that facilitate accumulation of capital by corporations.

FOCUS SECTION: IMITATION MEAT

Recent research indicates that most people in wealthy countries eat more meat than is healthy for them or the planet, and reducing meat consumption is one of the most important things that people can do to cool a warming planet. Insects provide a healthy, sustainable source of protein and are consumed many places in the world already (and in ways we often don't know about), but the yuck factor is a bit too high for most people to chomp on some crickets or mealworms. Two recent trends in artificial animal protein research and development are slightly more palatable, maybe. They are two slightly different interventions, using similar techniques: (a) imitation meats and (b) lab-grown meats. Imitation meat is made from plant protein. The most popular and available brand, Impossible Burger, is made from soybeans and potatoes, fats from various oilseeds and other binders to hold it together. What makes this glorified veggie burger taste like meat, however is the inclusion of an iron-containing compound called heme. It is found in all plants and animals, but the folks at Impossible Burger inserted heme from soy into genetically engineered yeast and fermented it to improve the taste. The second innovation is lab-grown meat, which takes cells from living animals and grows them in large bioreactors and then combines them with plant-based products in a patented process. The energy requirements for lab-grown meat are still quite high, making it an unviable alternative to meat in terms of mitigating climate change, although that may change as production is scaled up. Impossible Burger is widely available at fast-food and upscale restaurants alike. Lab-grown meat is too expensive and still in development to be readily available, although late in 2020, lab-grown chicken was approved for human consumption in Singapore. Mouat et al. (2019) argue that the imitation and lab-grown meat innovations are simply capitalizing on consumer anxieties and desire for sustainable alternatives.

SUMMARY

Protein is a necessary macronutrient for humans and livestock. Most available in animal tissues, but also in plants, such as legumes, it is expensive, environmentally costly and ethically questionable to produce it industrially in the form of animals. Like grains, meat production in wealthy countries requires massive subsidies from powerful nation-states. Dozens of environmental, social and human health issues are associated with protein production and consumption, and this chapter only touched on a few. Consumers, albeit with little success due to their relative lack of power in the food system, have pushed for alternatives or the elimination of animal consumption. Sustainable alternatives, in terms of what to eat and how and where to produce protein for human needs, abound in the wake of consumer demand for change. Locally produced meat, dairy and eggs have the capacity to provide economic and environmental benefits for rural communities and are emerging as viable alternatives, although they are limited in scope and scale due to policies that favor large-scale production. Lab-grown and imitation meats have the potential to revolutionize production methods but also reward powerful corporations for capitalizing on and profiting from ethics and the push for sustainability in new ways.

SUPPLEMENTAL MATERIALS

Key Terms and Concepts: albumin, aquaculture, aquaponics, blue revolution, bovine spongiform encephalitis, cold chain, confined animal feedlot operation (CAFO), Country of Origin Labeling, fish, geographic indications, integrated crop-livestock systems, lactase persistence, lactose, McDonaldization, meat, nutrient management, pasteurization, poultry, protected designation of origin, protectionism, rationalization, subsidiarity, territorialization

Explore More: ESRI – the company responsible for making much of the Geographic Information Systems software that geographers use to visualize data – has a Story-Maps series that allows users to post content about their interests. One such user created a series of interactive maps on livestock. Explore more here: https://storymaps.arcgis.com/stories/58ae71f58fd7418294f34c4f841895d8 *or* use Google to search "story maps farm animal planet"

www.seafoodwatch.org is a comprehensive website that allows users to search for best choices in seafood consumption.

Recommended Viewing: *Food Inc.* (2008), *Modern Meat* (2002), *Fast Food Nation* (2006), *Eating Animals* (2018), *Barbecue* (2018)

Discussion Prompt: Protein has become a controversial topic as the ethical treatment and climate change realities of animal production have surfaced in recent decades. What is your primary source of protein? What do you know about its production? Is it labeled in any way? Aim for 100–250 words. Be sure to correctly use at least *one concept* from this chapter *or* another chapter in your discussion post and write it in **bold** type.

Recipe: Every year in the spring and summer, my 14 chickens lay more eggs than we can possibly eat. I like to put them to use in a summery French dessert called clafoutis. It is simple and can be made with all local ingredients: eggs, cream, honey and any kind of fruit. You can find a recipe to get started here: www.epicurious.com/expert-advice/how-to-make-clafoutis-with-any-kind-of-fruit-article

REFERENCES

Cei, L., Defrancesco, E., & Stefani, G. (2018). From geographical indications to rural development: A review of the economic effects of European Union policy. *Sustainability, 10*(10), 1–21.

Dunn, K. (2019). Kaimangatanga: Maori perspectives on veganism and plant-based Kai. *Animal Studies Journal, 8*(1), 42–65.
Feagan, R. (2007). The place of food: Mapping out the "local" in local food systems. *Progress in Human Geography, 31*(1), 23–42.
Gerhart, A. (2017). Petri dishes of an archipelago: The ecological rubble of the Chilean salmon farming industry. *Journal of Political Ecology, 24*(1), 726–742
Harper, A. B. (2011). Vegans of color, racialized embodiment, and problematics of the "exotic." *Cultivating Food Justice: Race, Class, and Sustainability*, 221–238.
Mouat, M. J., Prince, R., & Roche, M. M. (2019). Making value out of ethics: The emerging economic geography of lab-grown meat and other animal-free food products. *Economic Geography*, 95(2), 136–158.
Nyéléni. (2007, February 23–27). *Forum for food sovereignty*. Proceedings held in Sélingué, Mali. https://ddd.uab.cat/pub/estudis/2007/180504/nyelenien_a2008.pdf?msclkid=6597f9e7aa8611ecb00d214773cfd729.
Ortiz-Gonzalo, D., Vaast, P., Oelofse, M., de Neergaard, A., Albrecht, A., & Rosenstock, T. S. (2017). Farm-scale greenhouse gas balances, hotspots and uncertainties in smallholder crop-livestock systems in Central Kenya. *Agriculture, Ecosystems & Environment, 248*, 58–70.
Peters, C. J., Picardy, J., Darrouzet-Nardi, A. F., Wilkins, J. L., Griffin, T. S., Fick, G. W., . . . Méndez, E. (2016). Carrying capacity of US agricultural land: Ten diet scenarios carrying capacity of US agricultural land: Ten diet scenarios. *Elementa: Science of the Anthropocene, 4*.
Ritzer, G. (2013). *The McDonaldization of society*. Sage.
Salatin, J. (2007). *Everything I want to do is illegal*. Polyface, Incorporated.
Wrangham, R. (2009). *Catching fire: How cooking made us human*. Basic Books.

CHAPTER 10

Tropical Commodities

Beverages, Fruits, Sugar and Spices

INTRODUCTION

In 2012, after an investigation into child labor in fair trade cotton, Fairtrade International, a German non-profit responsible for certifying fair trade farms, conceded that no certification system can guarantee that any product is free from child labor. The question of how confidently consumers could purchase fair trade products when the possibility of actually knowing the conditions of production is virtually nil led me to investigate fair trade and organic banana production in the Dominican Republic. From my research in the Dominican Republic, I learned that the mostly Haitian workers on small-hold farms demonstrated little to no knowledge of fair trade benefits or its workings. The few who did know about it clearly communicated that the benefit went to other people working in banana production. The workers I interviewed cited a lack of access to bathrooms, precarious working conditions without visas and lack of access to adequate food, housing and water and repeated exposures by aerial applications of insecticides and fungicides. All of these things are clear violations of the fair trade standards. Why and how this can happen has many causes, a few of which are explained at the end of the chapter, but this is a general pattern associated with tropical commodities. The longer the supply chain, the more obscure is the system. This often leads to control of the supply chain by a few powerful people at powerful points who are accountable to no one, which increases the likelihood of injustice, both in terms of labor and also the environment. Figure 10.1 shows Fair Trade Certified bananas from the Dominican Republic boxed and ready to be shipped to Belgium.

This chapter focuses on tropical commodities that are produced as luxury goods in former colonies for their former colonizers and/or consumers in wealthy nations. The chapter discusses broad categories of tropical fruits, tropical beverages, sugar and spices. The colonial origins of tropical commodities are discussed, including the use of slave labor to produce them, which built the empires of the Global North and fueled the Industrial Revolution. Then discussed are the geopolitical and political-economic aspects of sugar, cacao and coffee production and the labor implications of the trade in tropical fruits and spices.

DOI: 10.4324/9781003159438-13

FIGURE 10.1 Fair Trade Certified bananas

KEY CONCEPTS AND TERMS: TROPICAL COMMODITIES

Fruits

Beginning in the early 1800s, at the height of the European empires, many people in the northern hemisphere began consuming **tropical fruits** grown in British and other European colonies. Oranges at Christmastime were an iconic treat featured in literature of the time. Now fruits such as bananas, mango, pineapple, papaya, avocado and citrus of all kinds are commonplace, and their appearance in the grocery store at all times of the year is not questioned. In fact, their absence would be notable. Tropical fruits include a wide array of products from the ubiquitous and cheap banana to the more exotic and unusual starfruit and dragonfruit. Easily half the produce section of every supermarket in the northern hemisphere is filled with them. In the winter months, fruits that can be grown in temperate climates are sourced from tropics or subtropics in the southern hemisphere. The traffic in fruit, like most tropical commodities, is decidedly one way – to the northern hemisphere from the tropics. As already discussed, this is a pattern that follows from European colonialism and American neoimperialism. Much like quinoa, producers of valuable exports typically do not consume them, eating other, less valuable fruits and vegetables, such as plantains in banana-producing places.

Beverages

Tropical beverages are a valuable category of exports all their own. Coffee, tea and cacao are widely consumed in the northern hemisphere on a daily basis, and due to the addictive properties of caffeine, which is found in varying quantities in each, demand is stable and high. Coffee drinking originated in East Africa around 850 CE and was spread by the Islamic empire to Europe. Coffee is made from the fruit of a midsize shrub when the inner bean is dried and ground. Tea drinking originated in Asia nearly 5,000 years ago in today's China and Japan, and spread to the west via European imperialism. Tea is made from the dried and sometimes additionally processed leaves of *Camellia sinensis*, related to the popular flowering shrub. The cacao tree, from what chocolate is made, originated in Brazil and was domesticated by the Olmec people of Mexico more than 4,000 years ago. It was spread around the world by European imperialism after the 1500s. While most of the world consumes cacao in the form of chocolate, cacao was historically made into a drink and is making a contemporary comeback in various forms.

Sugar

Sugar is a generic name for sweet-tasting, soluble carbohydrates. It takes many forms, some of which are discussed in the Chapter 11. Sugar is divided into two groups: monosaccharides, such as **fructose** or sugar from fruits, and disaccharides, which are composed of sucrose and lactose. Sucrose is the sugar we commonly associate with table sugar, which historically is derived from sugarcane but now also sugar beets. Sugarcane was first domesticated in the South Pacific, and the first record of sugar was found in Indian texts around 400 BCE. Like all other tropical commodities, sugar consumption diffused through the empire: Islamic to European and then European, primarily Spanish, to the Americas. Like many tropical commodities, the production of sugar was largely fueled by slavery and continues to be grown with questionable labor practices. Brazil is the largest producer of sugarcane and also the largest per capita consumer of sugar, some of which comes in the form of ethanol, which is used as automobile fuel. India is the second largest producer of sugarcane and largest consumer by country. Various alcohols, such as rum, are also derived from sugarcane and are valuable exports from sugar-producing regions, such as the Caribbean.

Spices

Spices are derived from seeds, fruits, roots, bark or other plant materials and are used as a flavoring or coloring for foods. Consumers today regularly and often daily enjoy a wide variety of spices, including black pepper and cinnamon, which have their origins in Asia nearly 4,000 years ago. Other spices are perhaps used less commonly but are as familiar are vanilla, allspice, fennel, cloves, nutmeg, cardamom, cumin, turmeric and ginger, some of which are shown in Figure 10.2. Spices are different from herbs, which are the leaves or stems of a plant that are used to enhance food. Some plants, such as fenugreek or cilantro, produce both an herb (cilantro) and a spice (coriander) from the leaves. Trans-imperial trade in spice was central to the religious traditions, including mummification, and the medicinal and culinary customs of Asia and Africa for centuries. Long before the rise of European empires, traders trafficked in spices, primarily through the Egyptian port city of Alexandria. Spices were very valuable in the Middle Ages in Europe onward, due to their preservative and flavoring properties, as well as their use in healing at a time when neither refrigeration nor modern medicine were available.

FIGURE 10.2 Spices from India (https://commons.wikimedia.org/wiki/File:Indian_Spices.jpg)

FOCUS SECTION: VANILLA

Ever wondered why vanilla is so expensive? At more than US$20 for 4 ounces, vanilla is only less expensive than saffron, a spice made up of the stigma and styles of the saffron crocus. One of the more labor-intensive crops in the world, vanilla is the fruit of a vining orchid native to Mesoamerica. The earliest growers of vanilla were the pre-Columbian Maya and Aztec, who added vanilla to their chocolate drinks. European colonizers took the plant to Britain, where it was grown in greenhouses but would not fruit because, like most orchids, it evolved to only be pollinated by a single insect, in this case a bee native to Mexico. Hand pollination became the norm for vanilla production, adding another labor-intensive step to the already delicate business of harvesting and drying, curing, a process that takes about a year. Eighty percent of commercial vanilla today is grown in northern Madagascar. Indonesia, Uganda and Mexico also contribute to the global supply of vanilla. In 2017, Cyclone Enawo struck Madagascar, a country already struggling with years of drought, and the vanilla-producing region in the north was directly hit. Prices soared in the wake of the damage to vanilla production and with it the production of synthetic vanilla, which is virtually indistinguishable from the real thing. For this reason, vanilla is the only spice to be regulated by the US Food and Drug Administration. The vast majority of vanilla consumed in processed food is imitation vanilla, which is made from coal tar, wood pulp and petrochemicals. The production process is environmentally damaging, and manufacturers are seeking new methods as the demand continues to increase. Nearly all of the vanilla plants were destroyed in Madagascar, and it takes many years for a plant to mature.

CONCEPTUAL ENGAGEMENT: THE TROPICS AND COLONIZATION

The **tropics** compose a band of geographic space 23.26 degrees north and south latitude from the equator, or the region between the Tropic of Cancer (north) and the Tropic of Capricorn (south). This includes countries in northern South America, the Caribbean, central Africa, South and Southeast Asia and Oceania. In these environments, the sun is directly overhead for at least 1 day per year, which is not possible in other regions due to the axial tilt of the earth. As a result, the average temperatures in the tropics stay above 65°F (19°C). Plants that do not tolerate cold temperatures or freezing and often fruit after more than a year of growth thus evolved in these climates and cannot be grown for production outside of them without significant intervention. There are a few exceptions. Some varieties of tea grow in the subtropics, but not very well, and they are not amenable for large-scale production. The top tropical commodities in terms of value as exports are bananas, sugarcane, cacao and coffee. Brazil is a key player in the production of all of these commodities, followed by India. Central Africa, the Caribbean and Southeast Asia are also developing markets for these commodities and often have pre-existing trade relationships with wealthy countries.

Tropical commodities, like nearly every food we consume today, have roots in imperialism and colonialism, but the conditions of their production, including slavery, persist. The transatlantic slave trade moved slaves bought in Africa and sold to plantation owners in the stolen lands of the Americas where they worked to produce cotton and sugarcane. These two crops were highly valued in the Industrial Revolution and are extremely labor intensive to produce. Given that laborers were not paid anything, not even what their labor was worth, all the profit on the commodity went to the plantation owners and into the hands of the colonizing empire via taxation. The standard of living that many people of European descent enjoy in both the Americas and Europe was built literally on the backs of enslaved African labor. The legacies of land dispossession and slavery persist today, in the form of **premature death**, lower quality of life, decreased life chances and discrimination against African descended and Indigenous populations in North and South America (Gilmore, 2002). The ways of thinking that made slavery possible (i.e., the idea that Africans and Indigenous people were not fully human and therefore not entitled to rights) persist. It is no coincidence that the wealth generated by the cultivation and trade of tropical commodities has not accrued to people in these populations generally and that some continue to experience poverty and food insecurity. Solutions to these problems, as indicated earlier, are not to make food cheaper or more available, but to change the way of thinking about people that renders some worthy of rights and protection and others not, which is what made this whole system possible. Discussed in what follows is how tropical commodities continue to be enmeshed with exploitation of labor and geopolitical and political-economic strategies for dominance on a global scale.

FOCUS SECTION: PAPAYA

Tropical fruits are often considered part of a healthy diet, and the eating of several servings of fruit per day is encouraged by dieticians. It is also often considered part of an ethical diet, such as vegetarianism or veganism, that is focused on fruits and vegetables. For a consumer in a northern climate, this means eating tropical produce at some point in the year, not all of which is produced ethically or sustainably. Ian Cook (2004) provides a narrative tour of the papaya production supply chain, noting its association with colonization and slavery as well as an alternative to sugarcane production in the Caribbean. The spread of papaya in Central America and the Caribbean follows the movement of people, initially, the Carib and Arawak people and then the Spanish and Portuguese after colonization. He shows how the supply chain for papaya and many tropical commodities is controlled by Europeans or American ex-patriots and/or the exporters who control distribution. Like Freiburg's green beans, quality in a tropical fruit that exhibits wide variability is controlled by top-down mandates that are difficult for most producers to meet. Papaya are easy to grow but have unusual patterns that do not fit well with standardization: they are large herbaceous plants, like bananas, with male, female and hermaphrodite sexes that can change sex with the climate. Males pollinate while females produce fruits in the form of small berries, but the hermaphrodite fruits are what consumers want to eat. These trees are vulnerable to various viruses that deform the fruit or stop the ripening process. Consumers expect a uniform product, which requires intensive management of the viruses and the plantation as well as careful handling of the fragile ripe fruit. Freshly picked papaya oozes a latex that burns skin, so workers must wear gloves to handle the fruit. Women, who do much of the packing house work for tropical fruits, told Cook they had to take time off due to the latex burns and were not given adequate protection for their hands.

EXAMPLES

Sugar Geopolitics

Sugar was introduced to Cuba, land previously occupied by Indigenous Taino and Arawak people, by the Spanish in the 1500s, but it was not until the late 1700s that Cuba became a world leader in sugar production. A century later, slavery was gradually phased out and replaced by low-wage labor, largely composed of the descendants of former slaves and immigrants from Spain and China. The US began investing in Cuba's sugar production as a way to secure access to sugar in the late 1800s; in 1902, Milton Hershey built a large chocolate factory in Pennsylvania and directly invested in the construction of a sugar mill in Cuba. He purchased tens of thousands of acres of land to grow sugarcane as a way to vertically integrate production and to control the price he paid for sugar. Hershey sold the mill a few decades later to the Atlantic Sugar Company. The Cuban Revolution, which was fomented in part by American control of land and corporate dominance of production (and corruption under a dictator), ended the corporate control of sugar in Cuba, and it was nationalized under the new government. The Hershey sugar mill still exists, now as a popular tourist attraction demonstrating the multifaceted dimensions of food production infrastructure over time to change and adapt to agritourism (Salinas Chavez et al., 2018).

After the Cuban Revolution and a century of volatile sugar prices in the wake of restructuring, emancipation of slaves, the collapse of empires and revolutions, the US invested financially in the only tropical geography in the continental US: south Florida. The Everglades were transformed into a 280,000-hectare zone of agroindustrial production called the Everglades Agricultural Area. During both World War I and World War II, sugar was a vital and scarce commodity, due to a decline in production by a third. This led to rationing at home and on the war fronts, but sugar was considered an essential food for soldiers because it was believed to increase their endurance. After World War II, the US decided that sugar was a commodity necessary for **national security** and could not be sourced solely from conflict zones such as the Philippines or Cuba (Hollander, 2005). The centrality of sugar to national security justified state-based investments initially in land to grow sugarcane in Florida, Puerto Rico and Hawaii, but later in the form of subsidies for corn production for high-fructose corn syrup (HFCS). In between investments in the development of sugar beets led to a marketing "war" between regions (Midwest where sugar beets are grown and Florida) over dominance in the supply chain. Since the development of genetically modified corn, the share of cane and beet sugar consumed by Americans fell in comparison to the rise of HFCS. This is a rather extreme but illustrative example of how the importance of one commodity framed as vital to national security became a central focus of protectionist state and corporate investment in public-private partnerships to secure sugar.

CHOCOLATE ECONOMICS

Cacao is not chocolate without sugar, and Hershey's investment in off-shore sugar processing facilities in Cuba was a strategy of **vertical integration**, whereby one company controls more than one commodity or production process in a supply chain. Cacao is native to Central and South America and was domesticated by Indigenous people. Cacao was once drunk ceremonially by the Aztecs and Mayans, and cacao beans were a form of currency throughout Mesoamerica in the pre-Columbian eras. The countries with the largest production are in west Africa, with Ivory Coast the largest producer. Cacao has two main varietals, one of which, the *criollo*, is rare and prized for its delicate flavor. It comprises less than 10% of total production. Eighty percent of production is from *forastero* varieties, which produce a lower-quality but more affordable bean. Chocolate is now ubiquitous globally and requires multiple steps in processing to transform cacao into a form that most consumers recognize. Figure 10.3 details the many steps in cacao processing. The early labor- and time-intensive steps of fermenting and drying take place in the producing

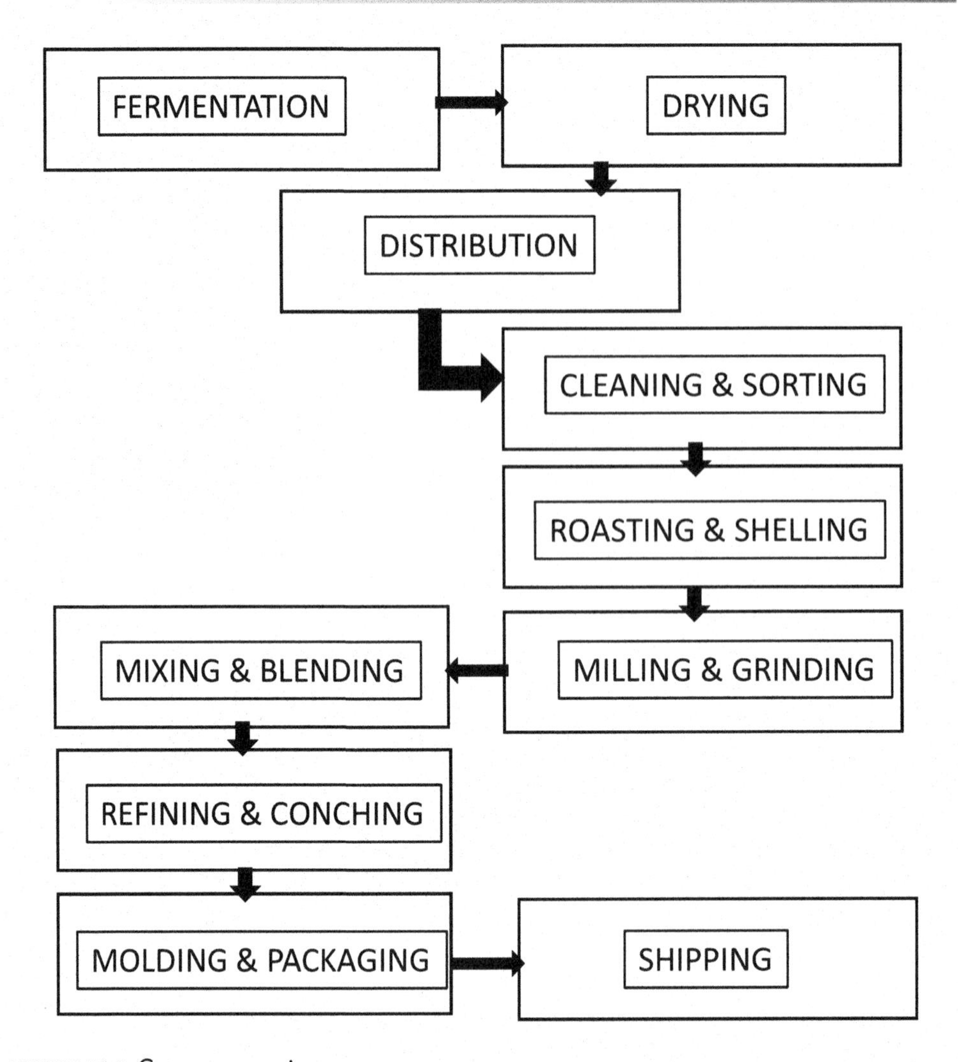

FIGURE 10.3 Cacao processing

country with low-wage labor. The steps after distribution often take place in large factories in developed countries where the product is consumed or then exported.

The vast majority of cacao is grown on millions of small-hold farms in the tropics, that, after the initial processing, sell to exporters. Chocolate manufacturing is controlled by three major corporations: Hershey's, Mars and Nestlé. Cacao is subject to volatility in price due to shifts in production, and small-scale farmers suffer greatly when they cannot make ends meet. In the late 20th century, a variety of news outlets reported the widespread abuse of children in cacao production, including slavery, sexual violence and torture. In response, many consumers began demanding more ethically produced chocolate, with the resulting dominance of fair trade certification making chocolate safe for capitalism again. Fair trade, however, is not a panacea, and small-hold farmers struggle to get a fair price and sustain their health, livelihood and environment. Due to demand outstripping supply for fair trade product, many of the large corporations have begun their own internal certification and labeling systems. Krauss and Barrientos (2021) argue that this shift to voluntary auditing for sustainability risks intensifying the power of corporations in the supply chain as well as re-establishing the conditions the fair trade certification set out to challenge.

COFFEE CHAINS

Coffee is an important global commodity, and consumers drink a couple billion cups each day. The US is the largest total consumer of coffee, while Norway is the top consumer per capita. (If you think you drink a lot of coffee, having drunk coffee until 11:30 p.m. at the Arctic Circle in July, I can confirm that Norwegians drink a lot of coffee.) Brazil is the top exporter followed by Vietnam and Indonesia. There are two general kinds of coffee. *Robusta* is a sturdy, hardy, high-yielding coffee species, originating in Ethiopia. It is primarily used as filler, instant coffee and espresso roasts. It has more caffeine and a stronger flavor. It composes about 30% of coffee cultivation and can be monocropped, or grown as a single crop. The other variety, *arabica*, was domesticated in Yemen and has a more delicate flavor. It composes about 70% of cultivation and grows best in **polycultures**, which are cropping systems in which multiple species grow. In the case of coffee, this often involves fruit trees growing in an overstory above the coffee plants.

Polyculture attempts to imitate natural ecosystems and were techniques of domestication and cultivation used by Indigenous people. In fact, European explorers mistook Indigenous cropping systems that included cacao and Brazil nut for virgin forests. Arabica coffee, which composes 70% of coffee cultivation and is the preferred product of consumers, prefers to be grown in shaded polycultures. "Shade-grown" coffee is a marketing tool that implies many things, including protecting rain forest biodiversity. That may be true, but as always, labeling claims require greater scrutiny. Arabica does not tolerate sun and must be grown with another plant taller than it, such as mango. This technically provides shade but is still considered a **monoculture**. Today coffee can be grown in commercial, traditional or rustic polycultures, depending on the degree to which native species are incorporated and the complexity of the system. Mango, lime and banana trees are often grown with coffee in a traditional polyculture with multiple levels of shade. Robusta can be grown at low altitudes in full sun, which does not affect its taste or productivity, which is why it is often used for cheaper coffee products.

The coffee supply chain is complex as shown in Figure 10.4. Coffee changes hands many times and crosses international boundaries before it arrives in the consumer's hand. Coffee, like many tropical commodities, is grown by small-holders, usually in some sort of cooperative group, or on plantations/estates. From there, coffee goes through a series of processing steps, often labor intensive. Exporters are the last stop in the producing country and typically control the market at a supply chain bottleneck. **Marketing boards** used to protect the price that growers would get but have been phased out as part of structural adjustment reforms. Fair trade as a replacement attempted to negotiate better prices for coffee growers and boost economic development. Wilson (2010) found that high levels of debt due to decades of low prices prevent efforts such as fair trade from succeeding. Coffee is then traded in various ways in the consuming country, roasted and sold through

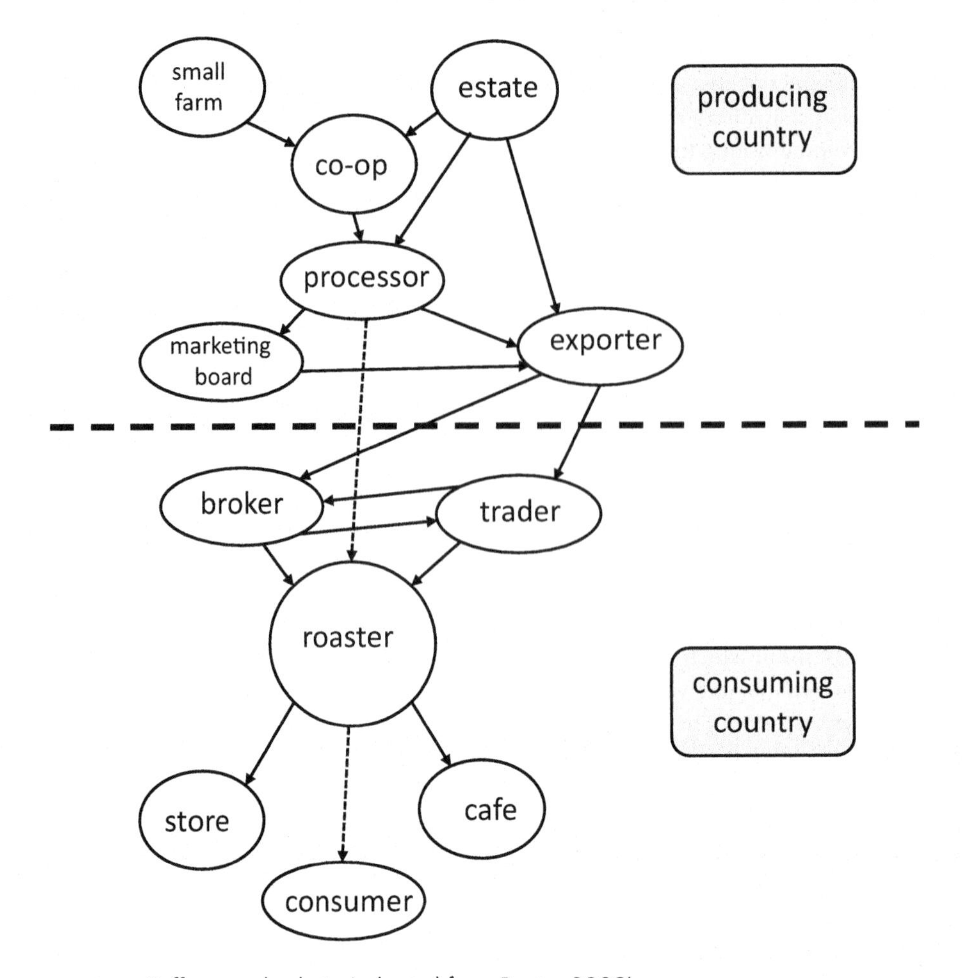

FIGURE 10.4 Coffee supply chain (adapted from Ponte, 2002)

retail outlets. Growers receive a tiny fraction of the money spent on coffee, the lion's share of which is reaped by retailers such as Starbucks (Ponte, 2002).

Multinational coffee corporations have led a push toward **shade grown** and **fair trade** coffee, although the vast majority of coffee is not labeled as such. Fair trade coffee emerged in the late 20th century in response to consumer concern about workers in the coffee industry and small-hold farmers who received so little money for their product they could not pay their production costs. This happens when levels of production increase to the point where prices fall. When producers use their land to grow something they cannot eat, like a normally valuable export such as coffee, market mechanisms break down, and food insecurity ensues. Some small-scale (and a few large-scale) coffee roasters have moved to a **direct trade** model to address the social and environmental concerns of coffee consumers.

Direct trade, also known as "relationship trade," is a form of product procurement that means that the processor or retailer buys coffee directly from farmers and exports/imports it themselves, thereby skipping all the brokers in the supply chain. This model has the potential to increase the share of money from coffee consumption going to producers as well as render the supply chain more legible. It is also a form of vertical integration, which in the hands of large and powerful corporations can lead to a downward pressure on prices paid to farmers. Direct trade emerged as a model when it became apparent that fair trade labeling of coffee had become something of an ethical greenwashing that did not improve the lives of workers or small-holders. Powerful actors in the supply chain still control the price of the vast majority of the coffee consumed in the world, however, and often small-scale growers cannot sell enough to make a living.

Focus Section: Direct Trade Spices

Spices have not received as much attention regarding their ethical labor practices or environmental impacts of production as other tropical commodities. Historically, spices were produced on plantations by low-paid or enslaved labor, sold by colonizing settlers to international trade companies and shipped back to the metropole of the empire along with other tropical commodities. Not much has changed in that scenario, including the working conditions. But now that is changing. Small-scale producers in tropical countries are partnering with companies in the US and UK to direct-market high-quality spices to consumers and high-end restaurants. Also known as the boutique spice trade, the development of new supply chain arrangements and relationships with producers has the potential to raise consumer awareness about spice quality, ethics in production practices and environmental health. Few consumers pay much attention to the quality of their spices, and most of us use spices that are years old from unknown origins due to a lack of labeling, governance and quality control for these products. Direct trade in spices has the potential to increase the price for producers, incentivize environmentally sensitive production practices and deliver a higher-quality product to consumers. For example, cardamom, shown in Figure 10.5, is a spice that is produced from a wild plant originating in India. It is the third most expensive spice in the world after saffron and vanilla. Today, the largest producer is Guatemala, and the Indigenous Q'eqchi' cultivate it in the mountain cloud forests of Alta Verapaz. Few farmers see profits from the sale of their cardamom, however, due to the long supply chain. A direct trade in spice could increase the quality, availability and income for farmers.

FAIR TRADE BANANAS

There are two general kinds of farms (shown at the center of Figure 10.6) for producing fair trade bananas for export in the Dominican Republic. The first is organized through banana associations that bring small-hold Dominican farmers, working farms less than 20 hectares, into collectives large enough to produce enough bananas each week to fill containers. The second is composed of large farms, also known as plantations or "hired labor organizations" greater than 60 hectares. Figure 10.6 shows the nature of the relationships between the actors in the fair trade production network, including the importer and exporters who are often involved in decision-making with the plantations and/or are part of the same company. The certifying bodies, shown at the bottom, inspect and

FIGURE 10.5 Cardamom flower (https://commons.wikimedia.org/wiki/File:Cardamom_flowers.jpg)

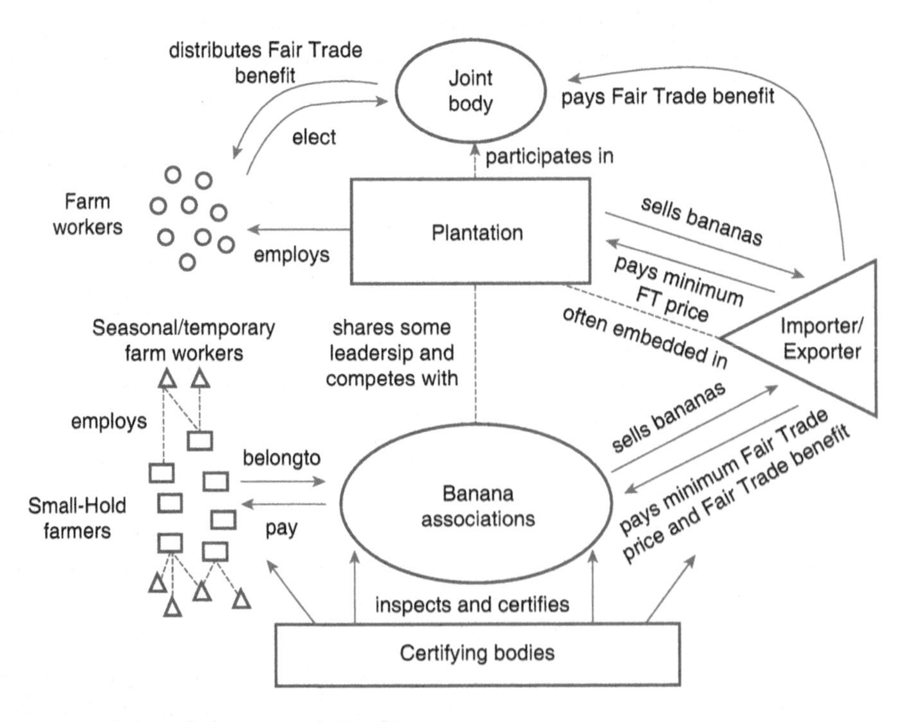

FIGURE 10.6 Fair trade banana relationships

audit every part of the supply chain to ensure compliance with the standards, some of which include protections for workers against aerial spraying of insecticides. Inspections are, however, not conducted without notice, which gives the organizations time to prepare for the audit. While conducting research, I witnessed multiple aerial applications of fungicides and insecticides while workers were in the fields (Trauger, 2014).

The figure also shows two additional relations, represented by dashed lines between the banana associations and the plantations, and between the small-hold farms and seasonal/temporary workers. The dashed line between the associations and plantations means that bananas are informally bought and sold between them to fulfill quotas, with plantations often augmenting the banana association's production to meet the quota for small-hold farmers. The dashed lines between the seasonal workers and the small-hold farmers indicate informal employment, often only for a few days each week. While small-hold farms are presumed to employ family labor, many only employ temporary workers. Fair trade was established to give price floors to farmers, so that they would be able to make a living even when global prices were low due to overproduction. It was also established to provide a premium (in the case of bananas, US$1 per box) that would go to banana workers on plantations to support food, education, medicine and housing. Seasonal or temporary workers are not allowed to be employed on small-hold farms, but these are typically the only kind of workers hired there in violation of the rules. They therefore do not receive the fair trade benefit, which is why, in the vignette that opened this chapter, small-hold banana workers on certified fair trade farms had no idea what it was.

SUMMARY

Tropical commodities have a long history of exploitation and dispossession associated with them. Historically, this meant enslavement of less powerful people and the colonization of Indigenous people's lands. Today, the continued consumption of tropical commodities includes their contemporary movement through long or extended supply chains that obscure the practices of production from the consumer. For some commodities this involves an unsustainable use of natural resources and in others, exploitation of labor that generates inequities. Some attempts to rectify the problems of tropical supply chains often reproduce the problems they set out to solve because they continue to be embedded in power relations shaped by capitalism and geopolitical strategies for dominance, which rely on the generation of inequality, threats to human health and the unsustainable use of resources.

SUPPLEMENTAL MATERIALS

Key Terms and Concepts: direct trade, fair trade, fructose, marketing boards, monoculture, national security, polyculture, premature death, shade grown, spices, tropical beverages, tropical fruits, tropics, vertical integration

Explore More: Return to the CIAT blog on crop origins from Chapter 3, and explore the website's many resources about tropical commodities. Choose one and find its origins and contemporary issues related to its production. https://blog.ciat.cgiar.org/origin-of-crops/

Recommended Viewing: *El Cacao: The Challenge of Fair Trade* (2015), *Caffeinated* (2015)

Discussion Prompt: What food do you eat that comes from the tropics? What is it, and where does it come from? Research its production and report on it here. Is it certified by any agency in any way? Aim for 100–250 words. Be sure to correctly use at least *one concept* from this chapter *or* another chapter in your discussion post and write it in **bold** type.

Recipe: Plantains are the larger, more savory cousin of bananas and are eaten in banana-growing regions of the Caribbean. Tostones are fried smashed plantains that are a delicious alternative to fried potatoes. They are simple and easy to make. A recipe is here: www.dominicancooking.com/301/tostones

REFERENCES

Cook, I. (2004). Follow the thing: Papaya. *Antipode, 36*(4), 642–664.

Gilmore, R. W. (2002). Fatal couplings of power and difference: Notes on racism and geography. *The Professional Geographer, 54*(1), 15–24.

Hollander, G. M. (2005). Securing sugar: National security discourse and the establishment of Florida's sugar-producing region. *Economic Geography, 81*(4), 339–358.

Krauss, J. E., & Barrientos, S. (2021). Fairtrade and beyond: Shifting dynamics in cocoa sustainability production networks. *Geoforum, 120*, 186–197.

Ponte, S. (2002). The latte revolution? Regulation, markets and consumption in the global coffee chain. *World Development, 30*(7), 1099–1122.

Salinas Chavez, E., Delgado Mesa, F. A., Henthorne, T. L., & Miller, M. M. (2018). The Hershey sugar mill in Cuba: From global industrial heritage to local sustainable tourism development. *Journal of Heritage Tourism, 13*(5), 426–439.

Trauger, A. (2014). Is bigger better? The small farm imaginary and fair trade banana production in the Dominican Republic. *Annals of the Association of American Geographers, 104*(5), 1082–1100.

Wilson, B. R. (2010). Indebted to fair trade? Coffee and crisis in Nicaragua. *Geoforum, 41*(1), 84–92.

CHAPTER 11

Not Food

Ingredients and Additives

INTRODUCTION

Several of my family members, including my mother and my aunt, have celiac disease, which is a lifelong autoimmune disease that damages the villi in the small intestine that help humans digest food. My daughter has also developed a gluten sensitivity that has yet to be diagnosed as celiac disease but which exhibits all the same symptoms. It developed suddenly a few days after Thanksgiving, which in American settler tradition is now a feast filled with gluten-containing foods. Until this moment, I had prided myself on cooking and serving whole foods for us and rarely let processed foods into my house, much less our mouths. It wasn't all kale and sweet potatoes all the time, but I made my own pizza crust, waffles, bread and her favorite burritos from scratch. But now, if I were to have any kind of sanity and she would have a normal life, we would have to make some adjustments.

Enter xanthan gum. A quick search of the goods in my pantry and freezer reveals xanthan gum in gluten-free burritos, pizza crust, crackers, bread, waffles, pasta, granola bars, candy, pancake mix, biscuit mix, brownie mix and so on. Anything that used to have wheat now has xanthan gum. Xanthan gum is a binder that does the work that wheat gluten does to make baked goods and others stick together and not fall apart. Xanthan gum derives its name from the species of bacteria used during the fermentation process involved in making it, *Xanthomonas campestris*. This is the same bacterium responsible for causing black rot to form on broccoli, cauliflower, and other leafy vegetables and is shown in Figure 11.1.

Most reactions to food additives are mixed: blissful ignorance of their origin and function, grateful acceptance of what they do to food and mild to extreme revulsion when the truth about their origin is revealed. This chapter provides an overview of synthetic and natural ingredients and additives used to enhance flavor, color or texture and the nutritional profile of food. While there is a dizzying array of additives, this chapter discusses four in detail: sweeteners, fats, preservatives and colorants The chapter explores threats to human and environmental health from food additives, public perception and misconceptions about the dangers of food additives and the governance of food substances that declare them safe.

DOI: 10.4324/9781003159438-14

FIGURE 11.1 Bacteria on cauliflower

KEY CONCEPTS AND TERMS: ADDITIVE FUNCTIONS

According to Eschliman (2015), food additives have four primary functions. The first is to maintain or improve *nutritional quality*. This is done through **fortifying**, or the addition of vitamins and minerals to foods such as enriched breads. Vitamin D is added to milk and orange juice, and iron is added to breakfast cereals and so on. This is in large part a response to both undernutrition and/or food insecurity in which the primary nutritional sources of these vitamins, fruits and vegetables are not available to people or are not widely consumed. Deficiencies in vital nutrients can cause serious and congenital health problems, such as birth defects. Riboflavin, for example, which is made from the *Ashbya gossypii* fungus, is part of the family of B vitamins, which also includes folic acid and niacin. They are now standard additives in many foods due to their capacity to prevent neural tube defects in fetuses. In fact, food manufacturers added folic acid to flour before the US Food and Drug Administration approved its use due to its high value to public health in preventing such defects. Processing also removes some vital nutrients from foods, and **enrichment** is a process by which those nutrients are returned through additives.

FOCUS SECTION: LYSINE PRICE-FIXING

Lysine is an essential amino acid necessary for protein synthesis in humans and other animals. It cannot be created by the body and must be consumed in the diet. Good sources of lysine are fish, eggs, cheese, other animal proteins and legumes. Animals, such as cows, that do not normally consume animal protein or chickens fed a "vegetarian diet" must have lysine supplementation in the feed. Even if they already get it from natural sources, such as alfalfa or clover in a pasture, or soybeans, livestock are frequently given it as a supplement to foster faster growth rates. It is vital to confined animal feedlot operations because those feeding systems rely on corn-based feeds, which limits the amount of plant-based lysine the livestock can consume. In 1996, lysine was the target of an industry-wide price-fixing scandal led by the major food and livestock feed corporation Archer Daniels Midland company. Its executives sought to reduce its cost through engaging in an illegal price-fixing that involved two other multinational companies, two in Japan and another in South Korea. **Price-fixing** is a verbal or written agreement (in this case, verbal, caught on tape) monopoly behavior between firms that stabilizes prices in a way that violates anti-trust laws and free-market principles.

The second function of food additives is to *aid in processing or preparation*. This includes things like emulsifiers, which help oils and vinegars stay in solution, or thickeners, such as agar, which is used to gel everything from ice cream to soup. Other aids in processing include a dizzying array of stabilizers, bleaching and leavening agents, compounds used in the maintenance of pH, moisture and product integrity, such as the prevention of caking or lumping. These include the commonly found baking soda, gelatin, propylene glycol and soy lecithin on food labels, which come from a variety of natural and synthetic origins, including plants, animals, minerals and petroleum. Other thickeners commonly used in food processing include carrageenan and alginate, which are also derived from seaweed. Agar and other seaweed derivatives have other uses in industry, medicine and agriculture as fertilizers. Seaweed is harvested from shallow ocean environments from farms in the coastal Africa countries, Philippines and other southeast Asian countries. Figure 11.2 shows a seaweed farm off the Swahili Coast on the island of Zanzibar. In Morocco, another large producer of seaweed, the demand for agar for vegetarian foods as a substitute for gelatin from animal sources is leading to unsustainable overharvesting and the destruction of marine environments (Callaway, 2015).

The third function of food additives is to *increase the appeal* of food, through texture, flavor or color. This may be a bit counterintuitive when the thing used to make them more appealing is essentially bug poop. Shellac is famous for making wood shiny, but it is also used to give candy a characteristic shine and is made from the secretions of the female lac bug. Also known as shellac, the resin is collected from trees in India and Thailand and processed into flakes, shown in Figure 11.3. The flakes are dissolved in alcohol and can be brushed onto foods, such as jellybeans or apples, giving them a shiny exterior that

FIGURE 11.2 Seaweed farm in Zanzibar (https://commons.wikimedia.org/wiki/File:Seaweed_farm_uroa_zanzibar.jpg)

FIGURE 11.3 Shellac (https://commons.wikimedia.org/wiki/File:Shellac_varities.png)

consumers find appealing. Other forms of improving the appeal of food may be a bit less disgusting, such as salt, sweeteners, fats and food coloring, all of which are discussed in more detail later in this chapter. Other additives such as caffeine may be used to stimulate or raise nervous activity in the consumer. Additives such as flavorings, like vanilla, are intended to improve or change the taste of food; colorings are meant to change or enhance the hue of foods that are changed through either processing or production; flavor enhancers, such as sugar or salt, appeal to tastes that humans especially crave. Which additive in which food makes it appealing, however, is a culturally defined process. For example, in parts of India, sugar is added to eggs instead of salt. In parts of Europe, salt is added to fruit to enhance the sweetness.

A fourth function of food additives is to *preserve food product quality or freshness*. The primary purpose of **preservatives** is to prevent **oxidation**, which causes spoilage due to exposure to oxygen, and shortens the shelf life of food products. Salt, sugar and alcohol have long been used as preservatives, until modern chemistry allowed scientists to develop more complicated solutions. Sodium nitrite and sulfites are two commonly used preservatives in meat and wine, respectively. BHT and BHA are two preservatives commonly used in processed and packaged foods as **antioxidants** for fats and oils. They effectively occupy the oxygen in a container so that it cannot act on any of the food products. They are used to protect a wide variety of products, including pharmaceuticals and cosmetics, and have a wide range of applications in industrial activities. They are derived from petroleum, and their use is restricted in some countries due to concerns about their carcinogenic potential.

In what follows, sweeteners, preservatives, fats and colorants are discussed in more detail, with a focus on their uses, safety and sources.

EXAMPLES

Sweeteners

Sweeteners are sweet-tasting compounds, some of which have enormous quantities of food energy and some of which have none. Humans evolved a taste for sweet things in their hunting and gathering days when concentrated, energy-dense carbohydrates were hard to find. They were either extremely seasonal, such as ripe fruits, and/or extremely dangerous to obtain, such as honey. Now, awash in cheap concentrated carbohydrates, such as corn syrup and sugar, humans need to find ways to limit them while satisfying cravings for them. Enter non-caloric sweeteners, such as aspartame, sucralose, erythritol and stevia, which are the most commonly sold today. Sucralose is hundreds of times sweeter than sugar, is sold under the brand name Splenda and was discovered in the process of making insecticides. It is heat stable, water soluble and has no effect on the human body, which makes it a popular choice for thousands of "diet" products today. Stevia is derived from a plant from the same name and is native to South America, shown in Figure 11.4, and used there for centuries medicinally and in cooking. Like sucralose, it is hundreds of times sweeter than sugar but has a bitter aftertaste, leading it to be blended with other sweeteners, such as erythritol, a sugar

FIGURE 11.4 Stevia plant (https://commons.wikimedia.org/wiki/File:Stevia_serrata_-_Flickr_-_aspidoscelis_(1).jpg)

alcohol. Other sweeteners approved for use include saccharin made from sodium or calcium salts and aspartame, a purely chemical compound discovered by accident, like many others, with an ill-advised in a chemistry lab, lick to the finger. Considerable controversy and back and forth on safety surrounds artificial sweeteners, some of it unfounded and some of it complicated by incomplete or inappropriate studies regarding their safety.

Controversy about safety to human health still surrounds a ubiquitous sweetener, high-fructose corn syrup (HFCS), which is found in thousands of food products ranging from cookies to wine coolers to condiments. HFCS was deemed an improvement to its thicker, less versatile cousin, plain old corn syrup. In addition to sweetening, it aids in browning, keeps foods moist, lowers the freezing point of food and extends shelf life. Due to the oversupply of corn, HFCS is both valuable to the food industry and affordable. The large quantities of cheap corn available as a raw material for additives led to the making of tens of pounds per capita each year. It should be no surprise then that nearly every processed food has some HFCS in it, unless it has made a commitment not to based on consumer fears and/or demand. A series of studies and reports in the early 2000s pointed to HFCS consumption and **metabolic disruption**, which is damage to the body's ability to make fuel from food, leading to obesity. The corn industry fought back and launched a series of advertisements about HFCS, deeming it safe "in moderation." How one is supposed to consume it in moderation, when it is in thousands of food products remains a question.

FATS

Another food product with a similarly sordid history in recent years is partially hydrogenated vegetable oil. Like corn, soybeans are an overproduced commodity that forms the basis for a myriad of food and industrial products, ranging from livestock food to lubricants and inks. A goal of early food manufacturers was a semi-solid fat that behaved like butter but was less expensive, with a lower saturated fat content, and was shelf stable at room temperature. In the early 1900s, French chemists invented a process called **hydrogenation**, which is the forcing of hydrogen atoms into an oil until it becomes a solid. For decades, Crisco, one of the early brands of hydrogenated oils, was king among home bakers and food manufacturers alike. That is until in the late 20th century, researchers discovered that trans fats that come from the partial hydrogenation process are very bad for heart health.

Food manufacturers scrambled to remove them and replace them with something less dangerous. Enter palm oil. Only recently available as a commercially viable alternative to other vegetable oils, palm oil is produced from the fruit of oil palm trees, endemic to west and southwest Africa, but naturalizes easily to tropical environments. It produces more oil per acre than many other oil-producing plants and is now widely cultivated throughout Southeast Asia. Palm oil is a naturally saturated fat without cholesterol and has many similar properties to partially hydrogenated vegetable oils. That does not make it a health food, however, and it comes at an enormous environmental and livelihood cost for the people and landscapes where it is produced. The vast majority of it comes from Malaysia, where endangered orangutans and Indigenous people are impacted by the large-scale palm plantations that have arisen in the past few decades. Many consumers are aware of the problems, but the product has become ubiquitous in processed foods, especially in foods labeled as vegan.

FOCUS SECTION: PALM OIL

In the past decade, palm oil has become the most produced and consumed vegetable oil globally. It is ubiquitous in processed products ranging from biscuits and ice cream to shampoos and vitamins. The social and environmental problems associated with oil palm production are legion: tropical deforestation, greenhouse gas emissions, loss of biodiversity and habitat loss for critically endangered plants and animals, soil erosion, water pollution, worker exploitation, Indigenous dispossession, land grabbing and eviction of Indigenous people from traditional hunting and gathering lands. As is typical of such large-scale production projects, the largest landowners and corporate owners benefit the most. Southeast Asia is home to the largest area of commercial planting of oil palm, followed by West Africa and Latin America. The industry, aware of its negative image, has launched a campaign for "sustainable" palm oil, although most consumers know very little about what that means in practice. A study conducted by Reardon Padfield and Salim (2019) indicated that most consumers in Malaysia, Singapore and the UK knew little about "sustainable palm oil," and while consumers across the three groups were interested in a sustainable product, only a small percentage were willing to pay more for a product that offered social and environmental benefits.

PRESERVATIVES

Like sugar, salt in the form of sodium chloride has a similarly fraught relationship with human health. A vital nutrient and one of the few preservatives available for millennia, its trade fueled empires that controlled the shores of shallow seas or were near vast underground deposits. Now, ubiquitous in human diets, and like sugar, with a taste for it that evolved as survival strategy, overconsumption wreaks havocs on our hearts and other vulnerable tissues. There are a multitude of different types of salt, based on the minerals associated with the environments the sodium chloride crystals grew in. Figure 11.5 shows sea salt manufacturing centers in New Zealand. Sea salt derives its color from minerals dissolved in the sea water. In addition to the table salt that most people know, these include the now popular Himalayan pink salt, Celtic gray sea salt, black salt from Cyprus and red sea salt from Hawaii. These salts take on some flavor from the minerals in them but are mostly popular due to aesthetics. Salt is salt, after all. Salt is extracted through one of three means: evaporation of sea water, leading to sea salt; mining of solid salt which is then crushed (and not included in food); and **solution mining**, which is water pumped into a salt deposit, which is flushed out and evaporated. This last method produces the majority of salt humans consume today, often fortified with an additional vital nutrient, iodine. Kosher salt, preferred by chefs in cooking, is iodine free, and its name refers to its use in brining or koshering meats.

FIGURE 11.5 Sea salt manufacturing, New Zealand (https://commons.wikimedia.org/wiki/File:Solar_salt._(8107355304).jpg)

Salt as a preservative for meat is as old as meat consumption itself, but in the 20th century, some new players were added. Really nasty bacteria, including salmonella and those that cause botulism, live in unrefrigerated, uncured meats. People eating meat in pre-refrigerator worlds salted, smoked and dried meat and fish to preserve and prevent the growth of bacteria. Salt works to dehydrate bacterial cells so they cannot grow and also has the nice effect of making certain meats, such as fish and pork, quite palatable and portable. While we still do this to a large extent, think bacon, ham, smoked salmon and so on, food manufacturers began adding something called **sodium nitrite** to preserved meats. It not only adds an extra layer of anti-bacterial protection, but it also makes meat pinkish in color, rather than the dull brown from traditional processing, thereby making it appear fresher and more appealing. Nitrites have a mixed reputation, like many food additives, given food scares, largely unfounded about their carcinogenic potential, with some researchers linking the consumption of cured meat to stomach cancer epidemiologically (Crowe et al., 2019).

In a modern food manufacturing plot twist, a substance once used as a preservative, alcohol, now requires the use of preservatives. The process of making wine produces sulfur dioxide compounds, also called sulfites, which are naturally anti-microbial. They also act as a preservative and prevent the oxidation of the grape juice into vinegar. The modern wine industry adds additional sulfites to wine in order to extend its shelf life and make it possible to drink wine that has aged for decades or more. Controversy surrounds this practice however, as sulfites are known to be the leading cause of the dreaded red wine headache, which may or may not be due to overconsumption and dehydration. More well-documented and potentially lethal is a sulfite allergy, which can build up over time and cause a severe reaction in an unsuspecting quaffer. Red wine has a higher concentration of naturally occurring sulfites, and as might be predictable, the added sulfite content to wines varies by processing, labeling scheme and geography. Organic wines in the European Union (EU) and the US strictly limit or ban the addition of sulfites. While the EU has a relatively low threshold for conventional wines, conventional wines in the US or Argentina have much higher limits. All wines, however, must carry a warning label that they contain sulfites, unless of course, they do not, in which case, you better drink it fast.

COLORING

Like salt and sugar, colorings are meant to make food more appealing, and in fact, consumers have come to expect certain foods to have certain colors even when they are not naturally occurring. Take for example, cheddar cheese. Cheese is normally white, but the region of England where cheddar originated (it has a **protected designation of origin**, by the way) was famous for its grass-fed Jersey cows that produced an orange-tinted, naturally beta-carotene (vitamin A) rich milk. When cows were moved inside barns through modernization, manufacturers added annatto color, made from the seeds of the tropical achiote tree, or paprika to mimic this coloration, leading consumers to believe cheddar is naturally orange. Early colorants were toxic, but modern chemistry has made many of them now safe for consumption, in spite of scary sounding complex names. Most go by their FDS numbers, like Red no. 5 and Yellow no. 40, which are made from a variety of

FIGURE 11.6 Cochineal beetle (https://commons.wikimedia.org/wiki/File:Pyrochroa_coccinea_102061567.jpg)

chemical compounds and are some of the most common artificial colorings. Blue no. 1, originally made from coal tar, is also an artificial color that has produced some controversary in recent years due to some studies linking it to hyperactivity and attention deficit disorder in children (McCann et al., 2007), causing some companies in the UK to replace Blue no. 1 with dyes made from species of cyanobacteria.

Colorings, like flavorings, are derived from either natural or artificial sources. Unlike flavorings, only one colorant can be designated as "natural" on food labels. Carmine is made from the dried body and eggs of the female cochineal (*Dactylopius coccus*) insect (shown in Figure 11.6) which lives on prickly pear (nopal) cactus in Mexico and South America. The largest exporter is currently Peru, shipping more than 4,000 tons annually. Carmine is used to color Campari liquor, yogurt and cosmetics. Carmine has the dubious distinction of being the only "natural color" allowed on food labels; all other colorants, regardless of their source, can only be listed as artificial. So, if you see the "natural color" label on your yogurt, now you know from where it comes.

SUMMARY

Food additives play a significant and growing role in making processed foods nutritious, shelf stable, long lasting and appealing. Because comprehensive food safety legislation does not require either the labeling of sources or the disclosure of potential harm, consumers, who have comparatively little power in the food system, are largely in the dark about the presence, functions and safety of food additives, many of which are from petrochemical origin. Those that are from so-called natural sources are often cloaked in secrecy about their origins because they may be repulsive to consumers. Long-term social, economic and environmental impacts associated with food additives are largely unknown because they are understudied, unless a food scare prompts investigation. Using the dubious wisdom of the "dose makes the poison," many view food additives as relatively innocuous due to the very small percentage they compose of the food product. In the absence of labeling by powerful corporations who control the supply chain, consumers do well to investigate for themselves what they are ingesting and its wider impacts on their bodies and the world.

SUPPLEMENTAL MATERIALS

Key Terms and Concepts: antioxidants, enrichment, fortify, hydrogenation, metabolic disruption, oxidation, preservatives, price-fixing, protected designation of origin, sodium nitrite, solution mining

Explore More: If your library has a copy, or access to the ebook, check out *Ingredients: A Visual Exploration of 75 Additives & 25 Food Products* by Dwight Eschliman and Steve Ettlinger. The book includes photographs of the additives, as well as detailed information about the source and history of their use.

Recommended Viewing: *That Sugar Film* (2014)

Discussion Prompt: In a processed food you typically consume, research one of the additives. Share what you learned about where it comes from and why it is used in your food product with the class. Be sure to read what others write, and do not write about the same thing. Aim for 100–200 words. Be sure to correctly use at least *one concept* from this chapter *or* another chapter in your discussion post and write it in **bold** type.

Recipe: I'm a type 1 diabetic and my daughter has celiac disease, so food in our house is definitely not typical. We try a lot of food that one of us ends up hating, and we rarely ever eat the same thing. Recently, however, I made some sugar-free sugar cookies with gluten-free flour and almond meal that we both loved. Xanthan gum and stevia for the win. Recipe here: www.truvia.com/recipes/sugar-cookies-recipe-with-truvia-sweet-complete-all-purpose-sweetener (I substituted a gluten-free flour blend and added a quarter cup of almond meal.)

REFERENCES

Callaway, E. (2015). Lab staple agar hit by seaweed shortage. *Nature News, 528* (7581), 171.

Crowe, W., Elliott, C. T., & Green, B. D. (2019). A review of the in vivo evidence investigating the role of nitrite exposure from processed meat consumption in the development of colorectal cancer. *Nutrients, 11*(11), 2673.

Eschliman, D. (2015). *Ingredients: A visual exploration of 75 additives & 25 food products.* Simon and Schuster.

McCann, D., Barrett, A., Cooper, A., Crumpler, D., Dalen, L., Grimshaw, K., . . . Stevenson, J. (2007). Food additives and hyperactive behaviour in 3-year-old and 8/9-year-old children in the community: A randomised, double-blinded, placebo-controlled trial. *The Lancet, 370*(9598), 1560–1567.

Reardon, K., Padfield, R., & Salim, H. K. (2019). "Consumers don't see tigers dying in palm oil plantations": A cross-cultural comparative study of UK, Ma-laysian and Singaporean consumer views of palm oil. *Asian Geographer, 36*(2), 117–141.

PART 4

Challenges in the Global Food System

CHAPTER 12

Labor

INTRODUCTION

My mom was a single parent and raised me and my brother on a small-scale subsistence farm in northern Minnesota. We had goats, chickens, sheep, horses, ducks and rabbits. It was heaven for a kid, and it shaped a great deal of my outlook on the world. When I was 13 years old, my mother married a corn and soybean farmer who had feeder calves in a feedlot in the southern part of Minnesota. Hilarity on my part did not ensue, but what did immediately transpire was me being drafted as part of my stepdad's "bean bar" crew. I'm not that old, but I do pre-date genetically modified crops, and my first paid job was to sit in one of the seats on the rig pictured in Figure 12.1 and spot spray sunflowers, cocklebur and burdock with herbicide in the hundreds of acres of my stepdad's soybean fields. The pay was not great, nor were the hours or working conditions, but my tan was slamming. And it was an improvement over the previous iteration of this job which was to dig weeds out by hand. The next year my stepdad bought genetically modified seeds and didn't need us to spot spray anymore. He just sprayed the entire field with Roundup, and so did all his neighbors.

This chapter covers a basic history of agricultural labor, governance of labor conditions, including social movements for fair working conditions, the use of migrant labor and the challenges to provide equity for all workers in the food system. Discussed is the perpetuation of the use of slave labor in the food system, with the addition of the restrictions on migration for farm workers and their labor-organizing efforts. Also discussed is the situation of restaurant and other food system workers during the historic shutdown in 2020 from the COVID-19 pandemic.

CONCEPTUAL ENGAGEMENT: LABOR PROCESS

My stepdad could afford to pay me and three other teenagers US$4 bucks an hour to spray his fields even though he made very little money on his crops for three reasons that are relevant to labor in agriculture everywhere. First, as teenagers our labor was not worth very much, since we had no skills beyond acting like a glorified spray nozzle. Second, we were technically underage and could not work anywhere else, but child labor is legal in agriculture. Last, the grain company that bought his grain would dock the price per bushel

DOI: 10.4324/9781003159438-16

FIGURE 12.1 A "bean bar" in Southern Minnesota (photo credit: Ole Hastad)

for every weed seed in the corn or soybeans. Purity is valuable when foods are turned into industrial products. This is part of the rationalization of raw materials in the food system that requires uniformity, predictability and standardization. In other sectors of the food system, rationalization takes the form of **assembly line** systems of production. These are widely used in meat packing and chicken processing. Each worker, often in close proximity to other workers, makes a particular cut and moves the bird down to the next worker. In both cases, the goal is to reduce the cost of labor and increase the value of the product.

Marx, nearly 200 years ago, outlined in one of the first analyses of wage labor in factories that industrial workers in capitalist systems are **alienated**, or "separated from" the condition of being human, from other workers, from the product and from the act of production. Workers exchange their time and the use of their bodies for a wage, often hourly, that reduces them to a piece of the system. They typically perform one action, often over and over, and rarely are involved in the creation of an entire product from start to finish. Alienation between workers maximizes their productivity, which increases profits for the factory owner. Depending on the place of work, talking is forbidden, breaks are limited, hours are long, women are forced into taking birth control and some workers are not allowed to leave. Repetitive motion injuries mean turnover is high, and workers have a difficult time increasing their wages. This is all part of a system to extract as much value from the laborer as possible. Any surplus, or profit, from the product is gained by paying workers less than their labor is worth or suppressing their capacity to earn higher wages.

Labor is a vital piece of every link in the food supply chain. The **labor process** involves the production of inputs, production of the raw material, manufacturing and packaging, distribution, transportation, marketing, retailing and preparation. Work in a typical supply chain is gender segregated and often racially segregated. Women tend to work in packing sheds while men tend to work in the fields. White people tend to own the business and do the public relations work, while people of color tend to be doing the heavy labor. Retail labor tends to be the most integrated, but it is typically low-wage work while White men tend to own the corporations that control the brands. Most labor at the production end of the supply chain is low-wage and low-skilled. In the case of the US, it is often located strategically in places where unions are effectively busted with right-to-work laws and/or with a high population of migrant workers who work without rights or protections.

In some places, as discussed later, this amounts to conditions of slavery where workers are trafficked and kept by force to work without pay. Small business owners, such as farmers and some retailers, are described as **self-exploiting** or their labor is effectively unpaid because they pay themselves from the profits. Most laws governing agricultural production allow children to work, leading to some horrific exploitation, such as that documented in cacao production. Women are often unpaid laborers on farms, but they are not necessarily self-exploiting since they may not legally own the farm or other means of production and are therefore not entitled to profits. The complexities of labor in a capitalist agricultural system can be distilled down to a single root cause however: the price consumers are willing and sometimes able to pay for food is quite low, and the costs of production are high enough that some employers resort to the exploitation and abuse of workers in the name of profit.

In what follows, the labor movement in the US is discussed as an example of the kinds of struggles that agricultural workers everywhere face. Following that, the relationship between labor and migration and the resulting human rights abuses that follow from the precarity of migrant worker status are discussed more broadly. Figure 12.2 shows women

FIGURE 12.2 Women working in a tomato packing shed in Pennsylvania

working in a tomato packing house on an organic farm in Pennsylvania. In general, women and men work in different environments on farms, with men working in the fields and women in the packing sheds. The majority of workers on this farm were migrants or Amish women, revealing the social cleavages around farm work and identity that shape exploitation. This is true even in the fair trade and organic supply chains, although some forms of labor have improved in these contexts. Last, slavery is an ongoing feature of some food supply chains and some examples are presented.

EXAMPLES

US Labor Movement

In the US, the contemporary **labor movement** began in the early 20th century in response to horrific working conditions, pitiful wages, the use of child labor in factories and inhumane hours. After bloody struggles in which many union activists were shot, some in the back, while fleeing police, the Fair Labor Standards Act was passed in 1938. The law included basic things that are now taken for granted, such as a legal minimum wage (US$.25/hour at the time), a legally binding 40-hour work week, and time and a half pay for any hour worked over 40 hours. In 1949, the Fair Labor Standards Act was amended to forbid the use of child labor, guarantee safe working conditions and provide paid sick leave (however limited) and some guarantees of sickness-related family leave protections. Agriculture was exempt from many of these provisions due to the seasonal, temporary and family-based nature of the work. For example, during harvest it may be necessary to work more than 40 hours per week and to get every hand involved, no matter how old.

However, agriculture was rapidly changing in the post–World War II era, and farms began to resemble factories and not family operations so much. American workers were also in short supply during World War II, leading to rationing and home gardening. After the war ended, returning GIs found good jobs in factories that were protected by the new laws and were not keen to work in the fields. The US therefore made a series of agreements with Mexico between 1942 and 1964 to import temporary contract labor. The program grew from 30,000 in 1958 to nearly 200,000 per year by 1951. Known as the Bracero Program, specific regulations protected workers, although enforcement of them was negligible since the US had no responsibility to non-citizens. The program established important migration and labor patterns, and its termination and the lack of a suitable replacement program fueled undocumented immigration. This left migrants with jobs in the US vulnerable to exploitation, wage theft and hazardous working conditions against which they ultimately rebelled.

The farm labor movement began in 1962 and was largely organized by Latino migrant laborers, including Cesar Chavez and Dolores Heurta. They organized farm workers into unions that would allow them to walk off the job to protest their working conditions. In 1965 the United Farm Workers led a 5-year strike by Filipino grape pickers, which included marches, hunger strike and boycotts to call attention to their working conditions and wages. The usual violence around labor organizing followed with lives lost, jobs forfeited to **scab labor** (people who will work and will not join the union) and people arrested. In 1970 the general strike ended and new contracts were signed, but given the lack of strong agricultural labor law, the struggle for fair wages, safe working conditions and other protections, such as from sexual assault for migrant workers continues.

The struggle for fair working conditions continues on a global scale. Today demands by laborers include a ban on the use of child labor; a minimum living wage; criminalization of **wage theft**, which takes place when the harvested food is "miscounted" or thrown out; and overtime pay so workers can choose, rather than be forced, to work longer hours for higher wages at harvest when demand is high. The Migrant and Seasonal Agricultural Worker Protection Act was passed in 1983 to establish better working conditions and to protect migrant workers from contractors, but it lacks enforcement. Migrants who wish to be legally employed in agriculture must file a H2-A non-immigrant foreign worker visa, which is expensive, time consuming and often is impractical for workers looking to work for a season or two.

LABOR AND MIGRATION

A **migrant laborer** is a worker who either moves location domestically in their country of birth or migrates beyond its borders for work. Most migrant workers practice what is called **circular migration**, in which they do not intend to stay in the location of their occupation but return to their home country at the end of the season. Globally, there are estimated to be 232 million migrant laborers, also known as **guest workers**. The countries with the largest populations include the US with more than 14 million, including an estimated 4–5 million undocumented workers. Europe as a whole and Saudi Arabia each host more than 5 million guest workers. Migrants work in multiple industries, with the primary being agriculture, followed by construction and domestic work as nannies and maids. As non-citizens, migrants are vulnerable to exploitation and abuse, especially if they are undocumented and fear deportation. Some are held as slaves. In the US there are about 2.5 million farm workers and about half are migrant workers, the vast majority of which are from Mexico and Central America in a pattern of economic activity established by the Bracero Program. Approximately half of those are estimated to be undocumented and are hired illegally by contractors and farm operators who take advantage of the lack of rights for workers (Holmes, 2013).

The Bracero Program and neocolonialism established a pattern of migration for farm labor in the US from Latin America. For the rest of the world, colonialism and imperialism shaped migration patterns. A *New York Times* article by Porter and Russel (2018) identified inter-region migration patterns for migrant workers as well as intra-region patterns. Three that shape food and agriculture globally are (a) the migration from Asia to Europe, (b) Asian migration within Asia and (c) European migration within Europe. Each pattern of migration has a similar story: one of uneven development and migration to a place of relatively higher development. The first pattern results from European colonialism in Asia, primarily British of South Asia. South Asian migration to the UK as part of the commonwealth resulted in the development of an entire new food industry that Britain is now famous for: curry shops. The second pattern of Asians within Asia includes a pattern of

Southeast Asian migration to other countries to work in seafood fishing and palm oil production. Many of these migrants are fleeing conflict or poverty and end up being trafficked into slavery. The last pattern includes eastern Europeans migrating to work in agricultural fields in western Europe after the fall of the Soviet Union. The collapse of many post-Soviet economies led to widespread unemployment and food insecurity, pushing migrants to travel west through the newly forming European Union (EU).

Much like the United States, farming in the European Union is still dominated by family run business, with a largely older, male base of owners. While the EU has the largest share of female farm owners, less than a third of farms are owned and operated by women. Farm labor is also largely provided by families, due in large part to subsidies through the Common Agricultural Policy that helps keep farm sizes small and hired labor requirements low (Garrone et al., 2019). While agriculture as a sector has declined in terms of employment, the number of people employed in agriculture remains high at around 10 million. As is the case in many places, farm labor is seasonal, sometimes part-time and peaks at certain times of the year (Roser, 2013).

That being said, fruit and vegetable production is heavily dependent on foreign nationals. Sweden, for example, employs seasonal migrant workers from Thailand to harvest berries, and the UK relies heavily on migrant workers from Bulgaria and Romania (European Union, 2021). Conflict and disaster, discussed in more detail in the next chapter, has significant impacts on labor in agriculture. In the case of conflict, refugees may be admitted as farm laborers from former colonies, and in terms of disasters, farm workers are often in short supply. In the case of the COVID-19 pandemic, a shortage of European workers led to improvements in worker pay and visas, many who were recruited on an emergency basis to harvest crops that would otherwise be lost. The influx of migrants to the EU also revealed deep inequalities in access to healthcare and rights (Neef, 2020).

The realities of agricultural worker exploitation are not just documented in the field, however. A report by the Southern Poverty Law Center found that workers in poultry processing plants endure repetitive motion injuries due to the expectation that they will work at the speed of the assembly line, making the same movements over and over again for hours. Like so many exemptions to labor laws, poultry processing plants are not governed by mandatory guidelines and are often self-regulating in response to market demands. While meat and poultry processing plants are subject to rigorous inspection schedules with regard to food safety to prevent *Escherichia coli* and salmonella outbreaks, they are not regulated for worker safety. And in the case of poultry processing in the US, the US Department of Agriculture has actually increased the speed of the processing line from 140 to 175 per minute in 2020, rather than establishing a speed that is safe for workers and consumers. Because the overwhelming majority of workers in poultry processing plants are migrants, they lack the formal protection of US law and are subject to firing and deportation if they resist or organize for their rights (Fritzche, 2013).

Contemporary social movements to challenge working conditions and to improve pay for farm workers take a variety of forms. The Coalition of Immokalee Workers launched a campaign to require one penny per pound of tomatoes sold to major retailers such as Publix and restaurants such as Taco Bell. Some retailers resisted the change, but those who implemented it have seen little change in prices. Other campaigns have demanded rights to such basic things as bathroom breaks and/or access to adequate bathroom facilities. Women workers have long reported widespread sexual violence, especially working in fields, and sexual assault is an omnipresent threat to women wherever they work in the food supply chain. Multiple agencies, organizations and scholars have documented the violence of pesticide exposures to farm workers and the adverse health consequences that follow. In short, farm worker organizations aim to call attention to the injustices that harm workers, environmental pollution and poverty. These conditions contribute to the food system contradiction of the two-class food system where the people who pick and process our food cannot afford to eat well. Social movements to change working conditions aim to harness consumer demand to make change that will allow for sustainability in terms of fairness and equity for laborers.

ORGANIC, FAIR TRADE AND LOCAL AGRICULTURE

The sustainability turn included an awareness of worker exploitation and a corresponding demand for better conditions for farm workers. Consumer demand drove the development of certifying agencies and standards for organic products in the US. As already discussed, consumers wanted more transparency about the conditions under which their food was grown, primarily in terms of what kinds of applications and inputs, such as pesticides, were used in growing them. Organic standards had very little to say about labor conditions on farms, and due to the various exemptions in labor law regarding agriculture, many questionable practices persist. During fieldwork for my dissertation in Pennsylvania, I volunteered on organic farms that were part of an organic marketing cooperative. One farm was run by a family with five kids, all of whom worked on their parents' farm in rural Pennsylvania, including the youngest of their homeschooled children. He was 4 years old at the time of my research, and he helped out in the packing shed applying the certified organic stickers on the boxes. This family sold their produce through a cooperative to some of the most expensive restaurants in the Washington, DC, area.

Another common form of labor on other farms in the cooperative was something called an **apprenticeship**. This is a form of low-wage labor used on small-scale organic farms that employs upper- and middle-class suburban college students during the summer break from school. The laborers were paid monthly stipends and room and board and worked to gain expertise in farming in the process. In some models, apprentices are responsible for growing, harvesting and marketing an entire crop through a season. Many of them were women aspiring to own their own farms one day, as was the woman cultivating in Figure 12.3. Unlike more conventional farms, certified organic or small-scale farms often employ women in a wider diversity of work, and a growing number are owned and operated by women. Many organic farms, especially the larger ones, employ seasonal and migrant workers from Latin America, sometimes illegally. The pay and working conditions were identical to the conventional farms they had worked on, according to the men I've interviewed, but they did note, tellingly, that on organic farms they were allowed to take breaks to use the bathroom.

FIGURE 12.3 Apprentice cultivating on a farm in Pennsylvania

FOCUS SECTION: COVID-19 AND THE RESTAURANT INDUSTRY

Restaurant industry labor is as precarious as other kinds of work in the food system. Service sector workers bear the brunt of economic downturns or other changes in demand for the luxury of eating out. The COVID-19 pandemic, however, was like nothing the industry had ever seen. The coronavirus outbreak in early 2020 triggered the shutting down of nearly all restaurants everywhere, some for a few weeks, others for so long they could not economically recover. When restaurants were allowed to reopen, they had to do so at limited capacity, some of them as low as 30% capacity, which meant fewer sales and worker layoffs. More than 2 million American jobs were lost in 2020 as the entire sector cratered. More than a hundred thousand restaurants in the US, no matter how much they are helped by stimulus packages, will not reopen when the pandemic passes because of the losses they sustained. According to the Centers for Disease Control and Prevention, eating out is one of the riskier behaviors to engage in during the pandemic, and in an effort to stay in business, many bars, restaurants and eating establishments stayed open out of desperation, leading them to be the source of community spread. For those who continue to stay employed, their working conditions have deteriorated, and with fewer patrons, tips have gone down, decreasing their pay.

SLAVERY AND AGRICULTURAL LABOR

In spite of being criminalized in every country in the world, slavery is still commonplace in certain industries. Nearly all contemporary slaves are migrants who are intercepted by traffickers as they attempt to escape intolerable situations in their home country. Slavery has been documented in the production of nearly every major category of foodstuff: sugarcane, cattle, coffee, cacao, nuts, corn, sunflowers, shrimp (prawns), fish, palm oil, sesame and beans. What these commodities all have in common is a long supply chain within which labor abuses can be hidden from view. Almost without exception, the product is produced in a country in the tropics and sold in a more developed economy. Some are staples, some are luxury goods, but all have a labor-intensive component to them, and the market price drives the exploitation of labor to extremes. Some people are kept as slaves through the mechanism of debt, also known as **debt slavery**, where fees and other bills associated with the cost of work amount to more than the worker is paid, compelling the worker to continue to work to pay the debt, which can never be paid. Other slaves are kept by more extreme measures such as captivity and brutality.

Shrimp, once a luxury food, is now the most commonly consumed, cheap seafood, and seeing them on all-you-can-buffets is commonplace. *The Guardian*, a British newspaper, undertook a years-long investigation into the use of slaves in the shrimp industry in Thailand. Workers are recruited from impoverished regions or conflict-ridden places throughout Southeast Asia through unscrupulous tactics, often promising work and high wages overseas. Once across a country border, the recruits are beaten, imprisoned and threatened. Men are sold onto ships that are at sea for months. Entire families are forced to work without pay in peeling sheds where the shrimp is processed, and workers report lack of protective gear, injuries and long working hours. A combination of corruption, complicity and a lack of regulation and enforcement allows the practice to continue and for shrimp to be sold to supermarkets in Europe, Asia and the US. Major retailers and restaurant chains count among those who sell shrimp caught and processed by modern-day slaves.

SUMMARY

Agricultural labor, long presumed to be the work of families and independent entrepreneurs, is rife with exploitation and worker abuse by powerful actors in the food system. The bottlenecks in the supply chain that concentrate power at the distribution points generate prices that benefit corporations but do little to support the work and livelihoods of farmers and farm workers from the least developed to the most developed economies in the world. The price that people are willing and able to pay for food is influenced by the power of major supply chain actors, but the low cost of food is also something to which most consumers have become accustomed and have little incentive to change. Small changes in wages and working conditions, however, make a big difference to workers, and most supply chains will survive increases to wages both in the poorly paid farm sector and the retail sector.

SUPPLEMENTAL MATERIALS

Key Terms and Concepts: alienation, apprenticeships, assembly line, circular migration, debt slavery, guest worker, labor movement, labor process, migrant laborer, scab labor, self-exploitation, wage theft

Explore More: Farmworker Justice is a non-profit organization seeking to improve working conditions for seasonal and migrant workers in the US. Explore the dramatically different rights for farm workers by state in an interactive map on their website. www.farmworkerjustice.org/general-map/

Recommended Viewing: *Food Chains* (2014), *East of Salinas* (2015), *The Harvest* (2010), *Dolores* (2017)

Discussion Prompt: Pick a food you regularly consume. Investigate who works to produce it at the farm level and where. Aim for 100–200 words. Be sure to correctly use at least *one concept* from this chapter *or* another chapter in your discussion post and write it in **bold** type.

Recipe: Fair Trade International admitted that no food can be guaranteed to be free of child labor and no food can be certified as slavery free, although some foods are more likely than others to be produced with slaves. Greens and garlic can be grown nearly everywhere by nearly everyone, and the only food you can be certain wasn't grown by slaves is food you've grown yourself. Investigate growing and cooking them.

REFERENCES

European Union. (2021). *Migrant seasonal workers in the European agricultural sector*. Retrieved May 3, 2021, from www.europarl.europa.eu/RegData/etudes/BRIE/2021/689347/EPRS_BRI(2021)689347_EN.pdf.

Fritzche, T. (2013). *Unsafe at these speeds*. Southern Poverty Law Center. Retrieved May 3, 2021, from www.splcenter.org/20130228/unsafe-these-speeds.

Garrone, M., Emmers, D., Olper, A., & Swinnen, J. (2019). Jobs and agricultural policy: Impact of the common agricultural policy on EU agricultural employment. *Food Policy*, *87*, 101744.

Holmes, S. (2013). *Fresh fruit, broken bodies: Migrant farmworkers in the United States* (Vol. 27). University of California Press.

Neef, A. (2020). Legal and social protection for migrant farm workers: Lessons from COVID-19. *Agriculture and Human Values*, *37*, 641–642.

Porter, E., & Russel, K. (2018). Migrants are on the rise around the world, and myths about them are shaping attitudes. *New York Times*. Retrieved May 8, 2021, from www.nytimes.com/interactive/2018/06/20/business/economy/immigration-economic-impact.html.

Roser, M. (2013). *Employment in agriculture*. Published online at OurWorldInData.org. Retrieved May 3, 2021, from https://ourworldindata.org/employment-in-agriculture.

CHAPTER 13

Conflict and Disasters

INTRODUCTION

The COVID-19 pandemic altered life for nearly every person on the planet in 2020. It has claimed millions of lives globally, and as of this writing, nearly a million people have died in the US alone. The vast majority of the pandemic's victims are elderly and vulnerable and disproportionately Black, Latino and Indigenous. The still unfolding disaster has devastated communities, closed educational institutions and collapsed entire industries. The COVID-19 pandemic exposed the weak points in the food supply chain in multiple, often brutal ways. COVID-19 outbreaks were severe in meat packing plants, throwing the meat supply chain into chaos, causing shortages nationwide. Similar outbreaks were documented in the apple industry in Washington and greenhouse facilities in Ontario, Canada, among others. These outbreaks impact low-income, often migrant workers who do not have sick pay or other forms of resilience to weather layoffs and medical bills, exacerbating the difficulties they already face. The restaurant industry, also deeply affected, responded in a variety of innovative ways, including moving dining outside, such as the pub in London shown in Figure 13.1, where it is safer, limiting seating capacity inside and mandating masks and social distancing. In other parts of the food supply chain, such flexibility is not possible or financially feasible or as open to the public eye.

Jackson (2006) writes that the internet can make disasters in faraway places loom large in the collective imagination, and the way in which people experience disasters closer to home is often invisible. As disasters go, the COVID-19 pandemic is unprecedented and exposed the vulnerable nodes in the food system's capacity to protect workers and provide reliable and safe foods. This chapter focuses on historic and contemporary challenges to food supply chains as a result of the natural disasters and geopolitical conflict. Topics include conflict with an emphasis on famine during wartime and food scarcity issues and disasters with a focus on supply chain disruptions due to pandemics, hurricanes and earthquakes.

KEY TERMS AND CONCEPTS: CONFLICT AND DISASTERS

Conflict is a state of ongoing hostility between one or more groups and often involves disputes over territory and resources. Territorial conflict has been associated with food

DOI: 10.4324/9781003159438-17

FIGURE 13.1 Outdoor dining in London during the COVID-19 pandemic (https://commons.wikimedia.org/wiki/File:We_Do_Take_Away_(50880130158).jpg)

insecurity for centuries, and the restriction of food is well documented as a weapon of war. Food insecurity is most heinously produced, to the point of intentional starvation by the withholding of food supplies to the opposition. This is a tactic of war that has been used everywhere in the world throughout history and is ongoing up until this very moment. Food security is also impacted by the destruction of fields and facilities that produce food, as well as the shortage of people to work in them, disrupting food supplies on the scale of the village to the scale of the world, as demonstrated by sugar shortages in World War II. Discussed in what follows are three cases of contemporary famine as a result of conflict that demonstrate the food security dynamics of war.

After nation-states become newly independent, civil war often breaks out in a struggle for control of the newly formed state's resources. South Sudan is the newest nation-state in the world after declaring independence from Sudan in 2011. Territorial disputes led to a civil war in 2013, which led to nearly half a million deaths and millions of refugees. Invading groups set fire to markets, stealing food and cattle and preventing farmers from working the land, leading to the increase in the price of food above what most people could pay. Blockades of food and aid were a common tactic used to starve the population in South Sudan. Other conflicts are ongoing in East Africa, including in Somalia and Ethiopia. Figure 13.2 shows a military protecting shipments of food into a war-torn area. Similarly, in Yemen, a rebel group took control of the capital in an effort to take control of the country. The Saudis and their allies bombed the country in defense of the prior government, destroying food and agriculture-based infrastructure. The Saudis gained control

FIGURE 13.2 Ethiopian soldiers help move food supplies through war-torn regions of Somalia in 2014 (https://commons.wikimedia.org/wiki/File:Ethiopian_soldiers,_belonging_to_the_African_Union_Mission_in_Somalia,_assist_in_the_movement_of_supplies_in_the_Bay_region_of_Somalia._AMISOM_Photo_-_Sabirr_Olad_(14224251760).jpg)

of the main port and effectively shut down any imports of food and aid, leading to widespread malnutrition and other humanitarian crises related to infectious disease and natural disasters. Nearly half the population of Yemen in 2020 was in critical need of food and aid to prevent starvation (National Geographic Society, 2021).

Like Yemen and South Sudan, Syria has been dealing with conflict-related food shortages, disrupted agricultural production and starvation as a weapon of war. For nearly a decade now, Syrians have been embroiled in a civil conflict with significant regional implications after several allied groups rebelled against dictator Bashar al-Assad. Government forces bombed critical infrastructure in rebel-controlled areas, including hospitals, markets and bakeries where people waited in line for daily food. Food shortages have been a way of life for so long, and aid and relief is restricted due to blockades and infrastructure damage so that some have taken to baking bread in underground bakeries to distribute food to the fighting and non-fighting populations alike (National Public Radio, 2013). In all conflicts, women and children compose the largest groups of civilian victims and refugees and also suffer acute food insecurity. As a non-fighting population, they are not prioritized for food. As refugees, they are non-citizens and not entitled to food, and their host states often do not provide it. International aid organizations such as the World Food Programme distribute emergency food aid, which is not nutritionally adequate over the long term, leading to stunting, nutritional diseases and vulnerability to infectious diseases.

Natural hazards are naturally occurring phenomena with negative short- or long-term consequences for human populations. They may be climatological (climate change), meteorological (hurricanes), geologic (earthquakes) or biological (viral pandemics) in origin. They often trigger **disasters**, which are measured by the scope and scale of human costs in terms of life, livelihood and financial losses. Natural disasters disrupt food supplies in various ways and have always been a major source of food insecurity. The scope and scale of natural disasters in the 21st century have changed, however, given the changing intensity of storms such as hurricanes and droughts due to climate change. Given that a small percentage of the population remains directly involved with food production in much of the industrial world, millions of people rely directly upon critical infrastructure such as grocery stores, which depend on roads, bridges, gasoline, trucks, labor, factories, farms and so on. The food supply chain is vulnerable to disaster, and where a weak point, such as concentration of processing in a few facilities, exists, one disaster can spell doom for the entire chain.

Hurricanes and cyclones that cause wind damage, flooding and foodborne illnesses and drought leading to crop failures and fires are the leading disasters that disrupt food production and supply. Figure 13.3 shows the deadliest disasters in 2020 that resulted in the loss of human life directly and also indirectly from the disruption of food and other vital life support systems, such as electrical grids and healthcare. Some systems, when they are resilient and redundant with other systems, recover relatively quickly. Other systems remain weakened and vulnerable to what Naomi Klein (2007) calls **disaster capitalism**. This is the intentional political and economic profiting from a disaster. In a general sense, capitalism as a process profits from disaster. Trees do not take on economic value until they are cut. To maximize the value from the cutting, trees are often clear cut, which constitutes a disaster for the forest and surrounding ecosystems. Disaster capitalism according to Klein takes the form of price gouging consumers for much needed food and water in times of scarcity or requiring people to pay for services such as electricity when they have no incomes. It also takes the form of governments taking advantage of the vulnerable community or population to establish ethically questionable policies that curtail human rights. This is also known as a state of emergency or what Agamben (2003) calls the **state of exception**, in which one or more crises are exploited to continue to roll back human rights in the name of order or security.

CONCEPTUAL ENGAGEMENT: SUPPLY CHAIN DISRUPTIONS

What disasters and conflict have in common is the disruption of vulnerable food supply chains. During the early days of the COVID-19 pandemic, consumers who had never experienced food shortages confronted barren store shelves. For some stores, this situation persisted for weeks and even months. **Supply chain disruption** happened in several ways. In the production sector, farms faced labor shortages and disruption due to border closures, limits on proximity in work spaces, illness, quarantines of exposed workers and disruption of manufactured inputs, such as fertilizers. In the processing sector, production was disrupted by a shortage of labor and through lockdown measures that forced workers

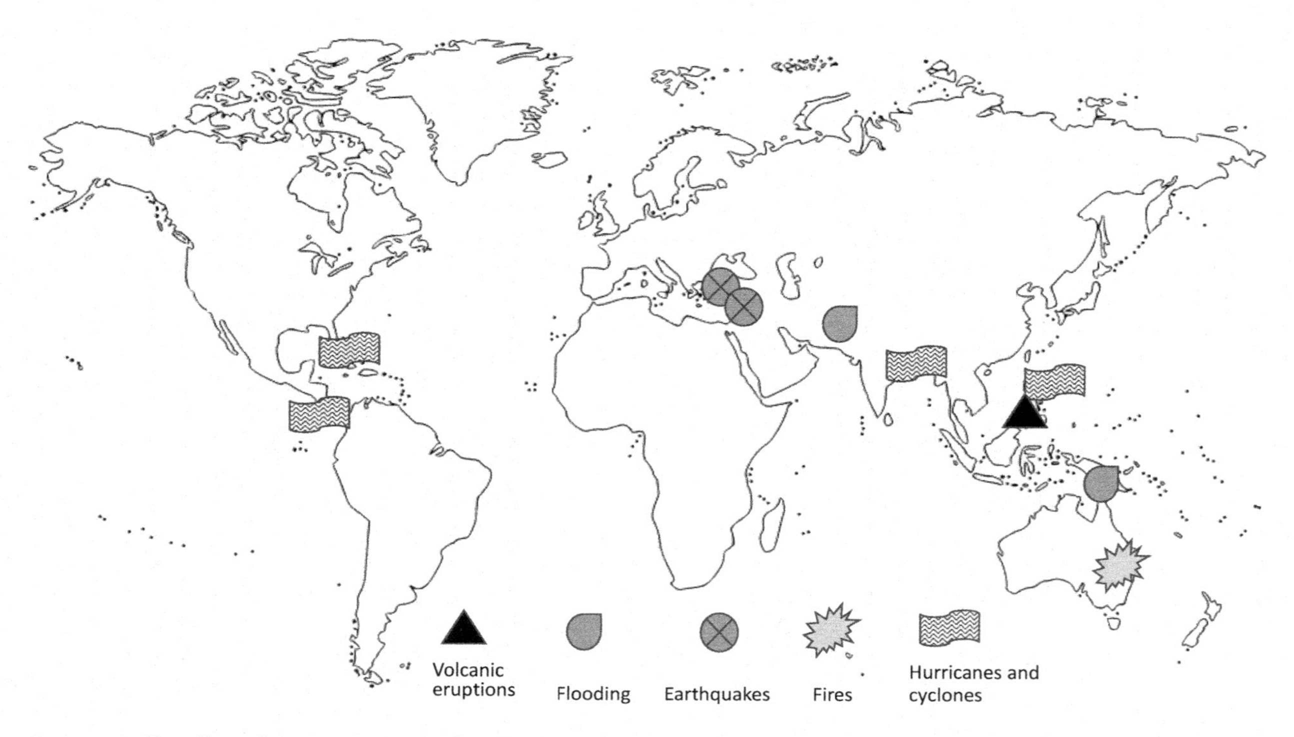

FIGURE 13.3 Deadliest disasters in 2020 (data source, *US News and World Report* 2020; https://www.usnews.com/news/best-countries/slideshows/here-are-10-of-the-deadliest-natural-disasters-in-2020?slide=12)

to stay home. Meat processing facilities in particular were hit the hardest, largely due to the labor intensity of the process. Transportation of food was disrupted largely through a combination of changes to air travel, border closures, social distancing measures at customs inspections, disruptions to shipping patterns, quarantining of ship workers and lockdown measures that forced workers into quarantine. Last, but not least, sudden changes in consumer behavior (i.e., eating at home for all or most meals) caused the restaurant industry to crater, and demand for groceries skyrocketed, leaving food retailers scrambling to meet the demand. The kinds of foods that people consume at a restaurant versus at home is quite distinct, and the dramatic shift in demand caused food shortages of some things and the loss of other kinds of food to waste.

Disasters like the pandemic exacerbate two major problems in the food system that together pose a contradiction. Disasters damage infrastructure, and any displaced or damage product must be thrown out without regard to its actual safety. The assumption is that if it was damaged or just not where it was supposed to be, it is potentially dangerous. While frustrating, damage to packaged foods can lead to foodborne illness, which is difficult to treat in a disaster situation, and also could lead to contaminated water and waterborne illnesses. This was a major contributor to the death toll after Hurricane Maria in Puerto Rico when people ate and drank contaminated food and water. When a disaster, such as an earthquake, disrupts a supply chain by damaging infrastructure and production facilities, food shortages are the result. As such, these two effects demonstrate one of the biggest weaknesses of the industrial food supply chain: it is dependent on long supply chains that are vulnerable in many places, especially because they could introduce pathogens when the safety of the food product cannot be controlled. Food that is out of place then causes a further shortage because it is lost or wasted.

EXAMPLES: SOLUTIONS TO SUPPLY CHAIN DISRUPTION

Food Aid

Emergency food aid is one of the first responses to disasters, and many organizations such as the World Food Programme have well-developed systems for delivering food aid to those in need. Figure 13.4 shows a large stockpile of food aid rations ready to be shipped at a moment's notice in the port of Miami, Florida, in the United States. Volunteers at a distribution point delivered food aid to people in Houston, Texas, after Hurricane Harvey flooded the city and destroyed infrastructure. Whether displaced by war or disaster, the greatest need of refugees is for food and water. The prevention of starvation and malnutrition is a primary goal of food aid organizations because these will trigger longer-term problems very quickly.

The goal of food aid is to provide culturally appropriate food that provides an adequate supply of energy even if it is not nutritionally balanced. Food rations are generally as simple and as durable as possible: a grain such as rice or corn, a concentrated energy source such as oil and a source of protein such as canned or dried fish or meat. Given that most refugees from destroyed homes from various sources will not have access to a kitchen to

FIGURE 13.4 Food aid rations stockpiled for an emergency (https://commons.wikimedia.org/wiki/File:20110826-FS-LSC-0063_-_Flickr_-_USDAgov.jpg)

cook food, vegetables are rare. Pregnant or nursing mothers are generally giving supplements. The ability to provision food takes time and energy in various forms, and food aid can be delayed due to damage to infrastructure and not everyone has the same capacity to arrive at a distribution point that must be large and centrally located, or small and less efficiently dispersed throughout a community.

Food Rationing

Communities such as those in Cuba that already have basic **food rationing** fare better in disasters. During the first three decades following its 1959 revolution, the Cuban government invested in industrial agriculture as a "project of modernity" through large-scale monocultural production on communal farms. In 1962, the Cuban government instituted a system called the *libreta*/ration book at very low, subsidized prices for food and other supplies. The government determines how much food each member of a household may acquire through the system, and this food is sold at local corner stores at state-subsidized prices that amount to about 10% of the actual cost. Much like food aid, the foods available are durable staples: sugar, salt, oil, dried milk, dried beans and rice. Animal protein is obtained in special meat stores and industrial products such as kerosene and matches at yet another store. These products are not sufficient to fulfill monthly household needs, and most Cubans buy additional food in state-run or private markets (Blue et al., 2021).

The ration system is an important safety net for families and provides some food security in times of crisis and provides a layer of resilience by already having a system in place to prevent starvation and malnutrition. The collapse of the Soviet Union at the end of the 1980s led to the rapid demise of Cuba's Soviet-subsidized industrial agriculture, imports of cheap food and a major restructuring of land tenure during a transitional period of deep austerity designated as the Special Period in Time of Peace. During the Special Period, power was out, buses did not run and the average Cuban lost 20 pounds. In the face of extreme shortages during the Special Period economic crisis of the 1990s, many urban residents responded to food shortages by participating in the large and growing urban agricultural movement.

Urban Gardening

Cuba's very successful experiment with urban food production, borne out of extreme necessity and supported in various ways by the Cuban state, has been an inspiration for communities suffering from food insecurity across the globe. Figure 13.5 shows a large garden outside a low-income community living in high-rises without access to adequate land. The garden feeds the community and is supported by international donors and

FIGURE 13.5 Urban garden in Havana, Cuba

investors. Another famous example is in the American city of Detroit, which responded to food insecurity in the Black community by turning empty lots into food-producing farms. The Detroit Black Community Food Security Network was born in 2006, and its farm lies on 7 acres in downtown Detroit. Led by Black residents and staffed largely by volunteers and funded by grants and donations, the farm produces more than 30 different fruits and vegetables that are sold at affordable prices through local markets (Yakini, 2010). Their goal is to have the African American community in Detroit meet its own food security needs through local participation in the food movement and for the community to have access to affordable, nutritious and culturally appropriate food. Poverty, disinvestment and uneven development are slowly unfolding disasters for the Black community, and local leadership organized to address these problems where people live. In other food deserts, such as in Atlanta and Seattle, volunteers, city planners, community leaders and local governments come together to grow a free food forest where anyone can pick fruits and vegetables. Modeled on permaculture design, the forest requires less labor to maintain and offers pecans, mushrooms, honey and blackberries in addition to the standard fare of tomatoes and kale in urban gardens.

Such urban interventions in response to ongoing or potential threats to food security are not new. The Aztecs built their capital city, Tenochtitlan, in present-day Mexico, on a marsh with islands dedicated to food production surrounding their city. Medieval European castles dedicated food production and storage within the city walls to prepare for conflict and disasters. The idea that modern cities can provide food for their residents as a way to bolster their local economies, preserve open space, mitigate climate change and prepare for disasters is, however, new. Urban planning during the rapid growth of cities in the late 19th and 20th centuries presumed that food production would take place outside the city limits in rural areas that were dedicated to agriculture. Modern city planners and decision-making considered agriculture, particularly animal husbandry, dirty, and this perception lowered property values. It forced many resilient forms of food production out of the city, making the poor vulnerable and making supply chains longer and more vulnerable.

Gardening, of course, remained a hobby in cities, but food animals were usually not allowed, and food-producing gardens were heavily regulated in terms of where they could be located. In the late 20th century, thanks to Cuba and other experiments, cities in the developed world began to see the value of **urban agriculture**, and in an ironic twist of fate, urban agriculture now raises property values. The largest proposed urban farm in the world in Paris was set to open in 2020 but was delayed due to the COVID-19 pandemic. It is designed to produce food organically for residents using an **aeroponic** method, which reduces demand for space, growing the crops vertically on raised lattices. While delayed in opening, demand for the produce, as in many places, was intensified during the pandemic, due to the disruption of other supply chains. Urban agriculture, while still in a nascent stage of development, shows promise for complementing the existing supply chain with locally produced food for residents, in a form of supply chain **redundancy.** This creates something called **resilience**, which can help people and communities withstand disasters and other ongoing threats to food security, such as poverty and supply chain vulnerabilities.

FOCUS SECTION: POLLINATORS

The loss of **pollinators** is a slowly unfolding natural disaster with unforeseeable consequences and for which we have no capacity to create redundant systems. Pollinators such as honeybees are vital to food production. They are responsible for the work that makes fruit possible, and some food crops, such as almonds, are completely dependent on honeybees for pollination. Like the vanilla orchid which would not make vanilla beans without its pollinator, our food system will dramatically change if we do not find a way to save honeybees and other pollinator populations. They are threatened as a group by exposure to pesticides, which are used in nearly every conventional agricultural system, and individual species have specific vulnerabilities to a combination of environmental change, parasites and habitat loss. Solutions to the problems posed to pollinators include an urgent need to change or reduce the use of pesticides and to restore habitat for native pollinators. Honeybees, which are not native but do the most important work in our food system, also benefit from the planting of organic crops that flower all season long (and not just in the spring) so that they can forage and make new queens for new colonies throughout the growing season. It is not one single threat but a combination of them working together to create a perfect storm (Watanabe, 1994).

SUMMARY

Disasters and conflict loom large in the collective imagination due to the ever-present threat of food insecurity and starvation that are often associated with them. These include supply chain disruptions due to the loss of agricultural land, laborers and the destruction of infrastructure or logistics that make it impossible to move food around. When the supply chain is controlled by a few manufacturers and distributors, these points also become weaknesses within the system if one or any of them fail. Urban agriculture and local markets have potential to provide resiliency and redundancy to food supply chains. In both cases, the promise of local and regional agriculture is underdeveloped for mitigating the threats to food security from conflict and disaster and deserves more attention, particularly following the supply chain disruptions during the COVID-19 pandemic. The food system dynamics of the slowly unfolding disaster of climate change is the subject of the following chapter.

SUPPLEMENTAL MATERIALS

Key Terms and Concepts: aeroponic, conflict, disaster capitalism, disasters, food rationing, natural hazards, pollinators, redundancy, resilience, state of exception, supply chain disruption, urban agriculture

Explore More: The National Oceanic and Atmospheric Administration collects and publishes data on natural hazards. Explore more with an interactive map of some of them. https://maps.ngdc.noaa.gov/viewers/hazards/
Recommended Viewing: *The Pollinators* (2019), *The Power of Community: How Cuba Survived Peak Oil* (2006)
Discussion Prompt: Do you, or did you during the early COVID-19 lockdown, experience changes to your diet, eating or food shopping patterns? If so, what were they? If not, why do you think that is? Aim for 100–200 words. Be sure to correctly use at least *one concept* from this chapter *or* another chapter in your discussion post and write it in **bold** type.
Recipe: Recall from Chapter 7 that jackfruit was a food of resistance for Sri Lankans in their fight for independence. Jackfruit is a nutritious and increasingly popular meat alternative. Look for it in well-stocked supermarkets, natural food stores and global farmers markets. A recipe for a jackfruit curry can be found here: www.veganricha.com/easy-jackfruit-curry/

REFERENCES

Agamben, G. (2003). *State of exception* (K. Attell, Trans.). University of Chicago Press.
Blue, S. A., Trauger, A., Kurtz, H., & Dittmer, J. (2021). Food sovereignty and property in Cuba and the United States. *The Journal of Peasant Studies*, 1–18.
Jackson, P. (2006). Thinking geographically. *Geography, 91*(3), 199–204.
Klein, N. (2007). *The shock doctrine: The rise of disaster capitalism*. Macmillan.
National Geographic Society. (2021). *Hunger and war*. Retrieved June 9, 2021, from www.nationalgeographic.org/article/hunger-and-war/
National Public Radio. (2013). *Jihadi fighters win hearts and minds by easing Syria's bread crisis*. Retrieved June 9, 2021, from www.npr.org/sections/thesalt/2013/01/18/169516308/as-syrian-rebels-reopen-bakeries-bread-crisis-starts-to-ease.
Watanabe, M. E. (1994). Pollination worries rise as honey bees decline. *Science, 265*(5176), 1170–1171.
Yakini, M. (2010). Undoing racism in the Detroit food system. *Michigan Citizen, 2*.

CHAPTER 14

Climate Change

INTRODUCTION

In 2004, I was doing fieldwork for my dissertation in geography in central Pennsylvania. My research questions were about how social networks help small-scale, organic farmers realize their economic or social goals, and my methodology was to help in the fields while we talked. During my fieldwork, Hurricane Ivan veered off the typical path of hurricanes and dumped 4–7 inches of rain on the mid-Atlantic in the middle of the harvest season. Figure 14.1 shows the storm tracks of 2004 and the unusual paths of many storms that entered the mid-Atlantic region of the US. The storm caused widespread flooding and 100s of millions of dollars in property damage. Dozens of counties were declared disaster areas, and there were millions of dollars in crop losses. On one farm, hit hard by Hurricane Ivan, I helped a farmer pick through her beet crop to find anything salvageable to sell to her organic market.

Her fellow cooperative members were in the same sinking economic boat, and she struggled to keep it together while we talked. As a small-scale food-producing farm, the crop was not valuable enough to be insured, the farm received no subsidies from the government and the income from feeding people organic vegetables was small on a good year. She talked for most of the interview about selling her farm and leaving agriculture. Such is the situation for many farmers who feed people in responsible ways. There is an enormous gap in the level of assistance that people who grow food receive, made ever more serious by the looming threat of climate change. In the last 15 or so years since Hurricane Ivan, nine hurricanes have dwarfed that storm in terms of size and damage, and there is no end in sight.

This chapter focuses on the unprecedented challenge presented to food systems by climate change. Elaborated on in this chapter are the contributions to climate change from the food system via greenhouse gas emissions and major threats to agriculture posed by climate change including natural disasters. The focus is on solutions such as the integration of agricultural and energy production systems in the form of biofuels, solar farms and carbon sequestration in place-specific ways. This chapter focuses especially on how power in the form of governance of food systems is a driving force in climate change, as well as a possibility for mitigation.

DOI: 10.4324/9781003159438-18

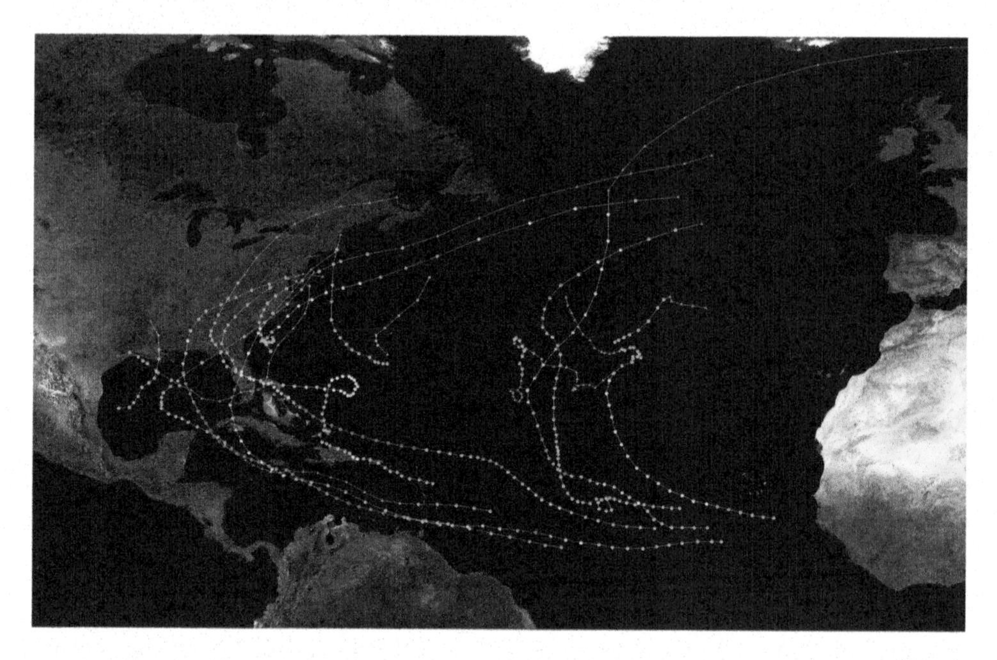

FIGURE 14.1 Hurricane tracks in 2004

KEY TERMS AND CONCEPTS: CLIMATE CHANGE

Once known as "global warming" due to the ominous uptick in global temperatures from the insulating effect of greenhouse gases on the earth's atmosphere since the Industrial Revolution, **climate change** has become the new moniker for global environmental change brought about by the changing composition of the atmosphere. Since the 1960s, scientists have observed a shift in the concentration of carbon dioxide (CO_2) in the atmosphere and a corresponding increase in global average temperatures. But since then, more sophisticated computer modeling made it possible to see that some places were cooling, such as the southeast US, while others were heating at unexpectedly rapid rates, such as the Arctic and the Himalayas. Storms, including winter storms and hurricanes, are getting stronger, droughts are getting longer and some places in monsoon climates receive catastrophic levels of rainfall and prolonged droughts in the same year. In short, the overall picture from changing the composition of the atmosphere is one of general, global change that often takes the form of extremes.

Computer modeling is still not up to the task of giving us very good projections in the future due to the data intensity of modeling the entire planet. Recall from Chapter 1 that everything is connected, and that the temperature of the Atlantic Ocean by Africa will affect the amount of rain that the coastal US will get from a hurricane. This is just one small geographic relationship that must be accounted for with models, and some relationships remain unknown. Therefore, scientific uncertainty about what is to come from climate change is not about whether it is happening, but how we can predict what is happening. The global temperature model shown in Figure 14.2 shows projections into 2100 and estimates based on what we can do to mitigate climate change. If nothing changes, warming will be catastrophic. If countries continue with agreements and accords already in place to reduce CO_2, there will be a greater than 2 degree warming, which will accelerate the extremes in drought and flooding that we already see at 1 degree of warming. Meeting the targets set to slow the change will require concerted effort. But the impact of the COVID-19 pandemic induced shifts in economic activity, notably reducing commuting. Whether those shifts might be maintained in the future remains to be seen.

The food system contributions to climate change are significant, and most experts estimate that food production and distribution contribute up to half of all CO_2 contributions. Figure 14.3 shows the percentage of contributions that come from food system activities. They come in the form of deforestation, which destroys carbon sinks and releases CO_2 through burning. Industrial farming methods use more energy in petroleum than they produce in calories that we consume, making it a net energy loss. Petroleum is used in cultivation, the creation of fertilizers and the production of pesticides. Livestock production also releases methane, a potent greenhouse gas, through animal waste and cattle alone make up half the total methane that the agricultural sector emits. Petroleum is used to transport food, as the vast majority of it moves from field to processing via tractor trailers. Petroleum is used as the base for all kinds of plastic packaging and to fuel processing plant machinery and to heat and cool buildings.

Food preservation also relies on fossil fuels, including coal-generated electricity to keep food frozen or cool and to heat and cool retail establishments. The last sector of fuel-intensive food system is food waste, which not only requires fuel but generates it in unproductive ways through decomposition in landfills. In many economically developing countries, foods are not necessary to be kept cool with a fossil fuel–intensive cold chain. Foods such as eggs do not need to be refrigerated, but industrial methods of handling them require it for food safety reasons. Similarly, fruits and vegetables sold in open-air local markets with

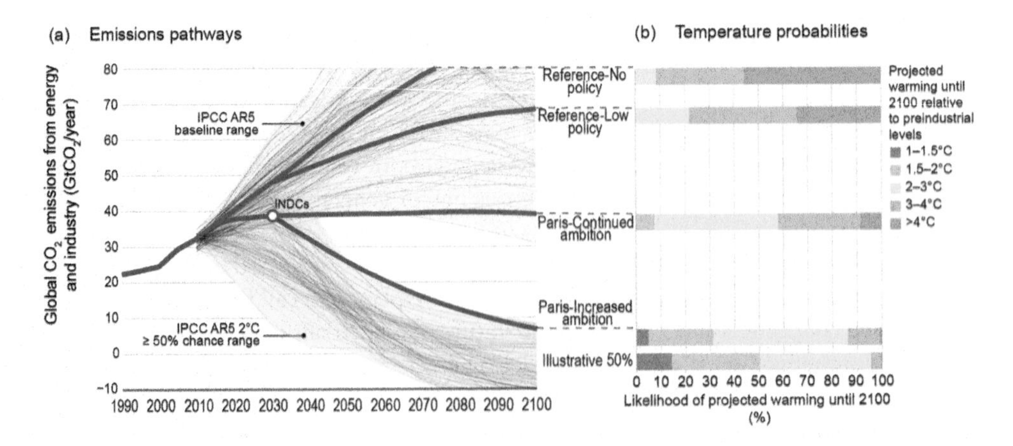

FIGURE 14.2 Climate change projections (https://commons.wikimedia.org/wiki/File:Global_CO2_emissions_and_probabilistic_temperature_outcomes_of_Paris.png)

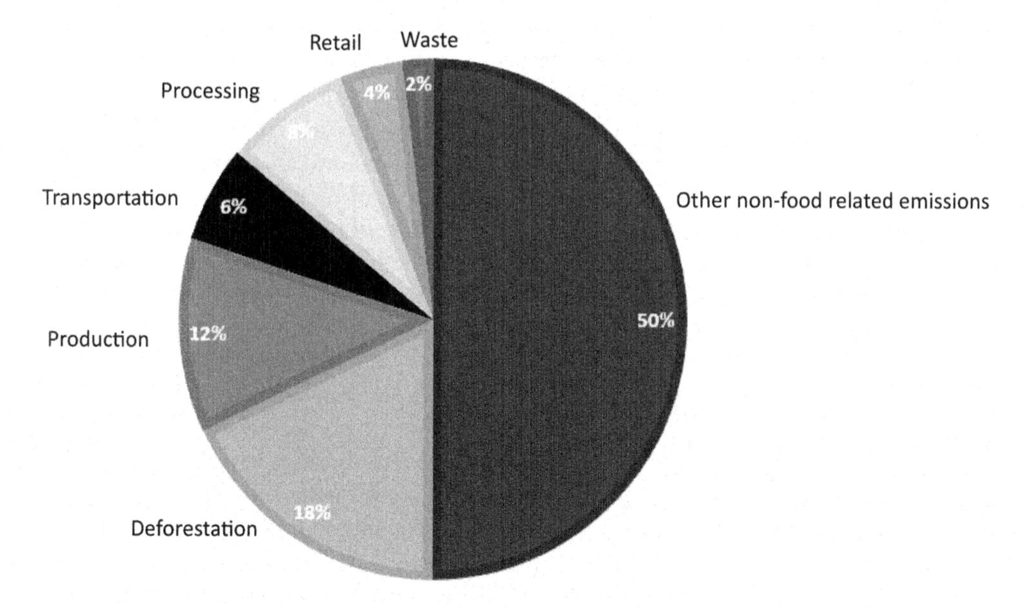

FIGURE 14.3 Food system contributions to CO_2 emissions (data source: La Via Campesina and GRAIN, 2014)

little packaging are not kept cool but are consumed the day they are harvested, requiring the work of women, typically to shop and prepare fresh food daily. Food waste is often fed to farm animals, and human and animal labor are the primary sources of energy fueling production. The world's wealthiest and most powerful nations contribute the most in terms of greenhouse gases, while the vast majority of the world's people will be affected the most, while they contribute to the problem the least.

CONCEPTUAL ENGAGEMENT: THREATS TO FOOD SYSTEMS FROM CLIMATE CHANGE

Researchers at Yale (Simmons, 2019) identified several major threats to agriculture posed by climate change. The first is that as *natural hazards*, such as fires and storms such as hurricanes become more dangerous and extreme, the damage to food production systems will become more extensive, widespread and permanent, increasing food insecurity. The world watched in horror in early 2020 when vast swaths, to the tune of 20 million acres, of Australia burned in an unprecedented natural disaster. As discussed in the previous chapter, natural disasters damage food production and distribution infrastructure, disabling entire supply chains in the process. In this scenario, the redundancy of local food systems may be helpful, but clearly not in the place that is utterly destroyed. In addition to losses to domestic food security due to the loss of livestock and farmland, Australia's large food export sector was damaged. Because of food shortages domestically, Australia will have to import food, but given the hit to its national income from exports, it may go into debt to do so. The Indigenous people of Australia who rely on natural ecosystems for some of their food security are also impacted in ways that will be difficult to recover from.

The wildfires in Australia were preceded by a decades-long drought that contributed to the scope and scale of the disaster. The second major threat to agriculture due to climate change is drought, which is a constant threat in arid and semi-arid areas. **Water scarcity**, or the lack of reliable access to fresh, potable water, is a way of life in such places. Drought exacerbates water scarcity, which may be produced in one of two ways. The first is lack of physical access, such as when wells go dry or water is polluted. The second, or economic water scarcity, is when the institutions to provide fresh water fail or when people lack the adequate resources to access it. Water scarcity exists in some form on every continent in the world, and in some delicate semi-arid environments, such as the Sahel, climate change is expanding desertification. In other more humid environments, such as Malawi, climate change is manifesting in catastrophic flooding, followed by a severe drought in 2016. Due to these back-to-back disasters, Malawi faced a severe shortage of grain, plunging its population into dramatic food insecurity. Normally food self-sufficient, Malawi produces enough grain to share with its neighbors in times of drought, demonstrating how climate change can challenge the most resilient systems (Kotir, 2011).

Last, climate change is doing what its name suggests, changing the climate on a microscale. This takes the observable form of seasonal shifts in **phenology**. Phenology is the study of seasons as relates to plant and animal life. Tree leaf and flowering dates over time indicate shifts in the overall climate. The emergence of spring peepers and the arrival of robins would be other indicators that together paint a picture of seasonal change. This has two effects on food production. The first is that warmer weather earlier extends the growing season, albeit in unpredictable ways. Early flowering of fruit trees can mean a better season, but a single frost happening at the normal time will devastate the crop, costing farm income and raising the price of food. The second effect is that the CO_2 that is fueling the change may adversely affect the food supply. CO_2 in some studies is shown to increase the food supply, but in a context of widespread food waste and loss, which is also a contributor to climate change, this may not be a good thing. Overproduction can also trigger new pests, and the extra supply could be damaged by new viruses and pathogens that a warmer climate makes possible.

Increased atmospheric CO_2 has been shown to increase the photosynthetic capacity of some plants and thus may contribute to increased food supply in some places. There are notable regional variations due to cropping patterns. The crops most likely to be impacted are wheat, rice and soybeans, which tend to be grown in the mid to high latitudes. They are staple food crops for billions of the world's population, so as long as these increased yields are not waste, this could be a boon for agriculture. The boon must be balanced against the fact that the climates in which those crops are grown is changing, however, shifting their optimal range north. Agricultural zones, or plant hardiness zones, are based on average last frost dates. Some tree and root crops also need a certain number of days below freezing, so this change in last date of freezing could expand the range of perennial crops northward (Cartier, 2021).

The way the industrial food system is structured and how it is governed is a driving force in climate change. Currently configured, food system governance increases the amount of greenhouse gases released through agriculture, and the will to change it is currently lacking. The UN Conference on Climate Change in 2021 did not focus on proven strategies to mitigate climate change, such as agroecology, and instead focused on energy-intensive technological innovations as solutions. Those with the most power to change the situation are doing the least, while those with the least power who are the most affected, such as the woman farmer in the example that opened this chapter, suffer the most. An important but overlooked aspect of climate change is the way in which certain groups of people are made disproportionately vulnerable (Bohle et al., 1994). This includes subsistence farmers in the Himalayas, which is known as the "Third Pole" due to the rapid pace of climate change in the region that is similar to the north and south poles.

Food system governance, if changed, can present an opportunity for **climate change mitigation** (reducing the driving forces and impacts of climate change) if leadership supports such a path. Recall from Chapter 1 that three primary actors govern the food system: states, civil society and markets.

The state, in the form of regulatory agencies and policy makers, is a key point with a goal of food safety and security. The state uses its power to structure the food system through subsidies and by providing support to citizens who need food and cannot purchase it. It facilitates partnerships with private entities such as corporations and non-profit organizations. The market sector is composed of businesses and non-profits and also supply-demand factors and, crucially, the continued reliance on food as a commodity. The third sector, civil society, is composed of communities and an emphasis on food as a common good, or something that is shared between individuals in reciprocal ways. In many ways, civil society is the weakest node in this system, but consumer power has driven many changes for consumer, worker and producer welfare in recent decades and may be a driving force for policy change to mitigate climate change. A few ways in which these three forces may use their combined power to craft solutions to mitigate climate change and reduce social vulnerability are discussed in the following sections.

EXAMPLES

Alternative Fuels

The primary force behind climate change is the emission of carbon dioxide from the burning of fossil fuels. Indigenous people in the Americas have been advocating for decades to "keep it in the ground" or to stop the extraction of fossil fuels altogether and replace them with alternative fuels. Indigenous people often are forced to live in close proximity to public lands where extraction takes place, and/or their homelands are crossed by polluting pipelines, as the recent Standing Rock standoff in North Dakota demonstrated to the public. Under pressure from consumers and other activists, via blockades, boycotts and disinvestment in fossil fuels, as some notable universities have recently done, states and corporations can be induced to invest in alternative fuels that can be used to fuel the food system.

One such alternative fuel is **biogas**, which takes the harmful methane emissions from livestock and anaerobically "digests" it in a system that produces a usable fuel for cooking, heating and lighting. These systems are currently in place on farms all over the world.

They are expensive to install and often rely on external sources of funding such as grants from non-profits and government support, but they quickly pay for themselves with savings on fuel. This will only become more attractive and effective as the cost of fossil fuel increases with scarcity. Biogas also shows promise to consume other kinds of agricultural waste, such as the straw from rice production, and of course, human waste. The systems show promise to be scaled up to the municipal level to address waste in a more widespread way. These systems require cooperation between states, non-profit entities and corporations to provide the funding as well as the involvement of individual farmers and citizens who stand to benefit, but they must adapt to change. Solar, wind and small-scale hydro power are other alternative fuels, but they do not show the same multiplier effect of removing a greenhouse gas and turning it into a fuel that biogas has the potential to do.

Rooftop Gardening

Much like urban gardening, rooftop gardening can take up unused space on a building, produce food and change the energy dynamics of the building. A Montreal, Quebec, company has taken the initiative to convert unused rooftop space into greenhouses in which local food can be grown throughout the year at high latitudes. Out-of-season temperate crops must either be grown in fossil-fuel intensive greenhouses or shipped from tropical or opposite-season climates in the southern hemisphere. In 2020, Lufa Farms opened the largest rooftop greenhouse in the world. Rooftop gardens are important carbon sinks where there would otherwise be a source of heating back to the atmosphere through something called **albedo**, or the reflectance of energy off light-colored materials back to the atmosphere. Roofs of buildings are painted white, of course, to avoid absorbing the energy from the sun and increasing cooling costs. A greenhouse on the roof doubles this investment by using solar energy to heat the greenhouse, reducing the fossil-fuel demand, and also absorbing the heat from the building that would otherwise be lost to the atmosphere. Add to these savings the fact that food can be grown and distributed locally without the need to ship produce across the planet. This was a project led by private industry, supported by the government and welcome, especially during the pandemic, by the public, who tripled the purchases when other supply chains failed.

Grassland Restoration

Carbon sequestration is the process of returning carbon to natural systems and cycles. Tree planting has been a popular strategy of mitigation, but Wes Jackson (2002) believes that perennial crops may be even better sinks for carbon. For decades, he has patiently been restoring native prairie lands and studying their potential as a carbon sink at the Land Institute in Kansas. With a PhD in genetics and from a family-owned farm in Kansas, Jackson realized that the key to mitigating climate change was to re-invent agriculture. To him, perennializing crops was a central piece of the puzzle, and he focused on creating perennial wheat and sunflowers that could be grown in a system that mimics a prairie ecosystem. The environmental knockoffs here include removing land from annual cultivation that erodes the soil and its carbon-sinking capacity, reduces the fossil-fuel demand for agricultural production and increases the carbon-sinking capacity of an underground root network.

The roots of a tree are similar in size and extent as the overstory in most cases, giving roughly the same carbon-sequestering capacity above ground as below. Trees also take up a lot of space that cannot be used for other things. Grasslands, however, can support crops and pasture livestock as well as host windmills and **solar farms**, or clusters of solar panels generating electricity, at the same time. Livestock such as bison, shown in Figure 14.4, are key to maintaining the biodiversity of a prairie and produce a low-fat, high omega–fat meat. Grasslands, when mature, have up to 10 times the biomass beneath the surface, which means that the enormous carbon-sequestering capacity of grasslands is actually under the soil, not above it. The conversion of croplands back to grasslands will take a concerted effort by individuals and organizations, who follow the leadership of Indigenous tribes already doing this. Changing the subsidy system for commodities is an enormous challenge but a significant one for combating climate change.

In a general sense, local and regional food systems offer a unique solution to the threats of climate change. As with everything, there are trade-offs. A local system has the potential to reduce reliance on fossil fuels as long as there is peak efficiency and/or alternative fuels are used within the system. Industrial methods produce fewer energy calories than they consume, so reducing their use is a good place to start. Most small-scale farming systems rely on human and animal labor and inputs, which helps. Some would argue that

FIGURE 14.4 Bison grazing on tall-grass prairie (https://commons.wikimedia.org/wiki/File:Tallgrass_Prairie_Nature_Preserve_in_Osage_County.jpg)

the transportation cost in terms of fossil fuels per unit of food is higher in a local food system, which just means that local food systems need to become more efficient. Local and regional food systems also have the capacity to provide resilience and redundancy to longer supply chains in times of disaster, but they too are vulnerable should disaster strike and a population is dependent on them.

Food Hubs

Therefore, we need local *and* regional systems to provide layers of protection in times of disaster and as the climate changes, potentially causing crop failures. Recent decades have shown enormous growth in local and regional **intermediated sales**, which means that farmers sell their products not directly to consumers but into an aggregated market made up of other small farmers, like a cooperative or a wholesale market. The food is sometimes sold through something called a **food hub**, which is like a farmers market but on a larger scale and with a wider reach than just individual consumers. Food hubs allow farmers to focus on producing a larger volume of a few crops instead of many and frees them from the labor of marketing their produce as well as growing it. Soubry et al. (2020) identify this as agriculture operating along *desire lines* as a way to shift food production toward social goals and increasing adaptive capacity for communities to face climate change.

While political activism is key to changing the system that furthers climate change, it bears repeating that consumers have the power to buy foods that support their ideals and shift the food supply to follow demand. Many people assume that a diet for a warming planet will include bugs and kale. That may be true if you are into that kind of thing, but it does not mean that we cannot eat the things we enjoy. It does mean shifting our consumption patterns to more sustainable sources, such as bison instead of beef, or grass-fed livestock products instead of confined feedlot livestock. It may mean buying local fruits and vegetables in the summer. It could mean preserving some of those by freezing or canning to avoid buying out-of-season produce. It may mean reducing coffee or chocolate or replacing it with more sustainable and ethical sources. It may mean eating more peaches than pineapple at high latitudes. Small shifts matter; all of these products are already available, and increased demand will only increase their supply. The locavore movement has shown us many times and in many ways that eating better means eating better for everyone.

SUMMARY

Climate change is a slowly unfolding disaster that is both produced through the industrial food system and also threatens the global food supply. The industrial food system relies on petroleum, and while the majority of the world's least powerful people do not contribute to the production of climate change, they are the most vulnerable to it. This includes disruptions to production, water supplies and an increase in natural disasters. The opportunities to mitigate climate change must be collectively met through changes in food system governance and cooperation between powerful governing actors such as states and corporations, who thus far have been unwilling to make big changes. These include synergistic activities that bring together a reduction in greenhouse gas production, mitigation of the impacts of the food system on the climate and regenerative carbon-neutral agriculture and food distribution systems. The hows and wheres of such future food systems and how to achieve them politically and economically are the subject of the next chapter.

SUPPLEMENTAL MATERIALS

Key Terms and Concepts: albedo, biogas, carbon sequestration, climate change, climate change mitigation, food hub, intermediated sales, phenology, solar farms, water scarcity

Explore More: The Climate Action Tracker evaluates countries with respect to their efforts to mitigate climate change. Explore more here: https://climateactiontracker.org/

Recommended Viewing: *Seeds of Time* (2013), *Kiss the Ground* (2020), *The Ants and the Grasshopper* (2020) found at: https://vimeo.com/396089404 Password: KTQprivate

Discussion Prompt: The existence of our changing global climate is not debated by scientists. How quickly the climate will change and where the greatest effects will be felt *is* quite uncertain, however. If you are a gardener or come from a farming family, this might be changing planting dates. If you survived a recent hurricane or other severe weather, this may be in the form of devastation from intense storms. If you live mostly in a virtual world, these may be things you see in the news or on TV. In 50–100 words identify and describe one way you see or feel evidence of climate change in the food system. Be creative and be sure to "read the landscape" either in the built or natural environment. Be sure to correctly use at least *one concept* from this chapter in your discussion post and to write it in **bold** font.

Recipe: Bison is an emerging sustainable source of protein that does the ecological work of mitigating climate change and reduces the social vulnerability of Indigenous people in the grasslands of North America, especially those who are trying to keep petroleum in the ground. Bison is also one of the healthiest meats you can eat, especially if raised on grasslands. There are many options to purchase online and look for Native-owned businesses raising bison on pasture. A recipe for a burger can be found here: https://healthyrecipesblogs.com/bison-burger-recipe/ Sub Impossible Burger for a vegetarian option.

REFERENCES

Bohle, H. G., Downing, T. E., & Watts, M. J. (1994). Climate change and social vulnerability: Toward a sociology and geography of food insecurity. *Global Environmental Change*, *4*(1), 37–48.

Cartier, K. (2021). *Climate change uproots global agriculture*. Retrieved June 11, 2021, from https://eos.org/features/climate-change-uproots-global-agriculture.

Jackson, W. (2002). Natural systems agriculture: A truly radical alternative. *Agriculture, Ecosystems & Environment*, *88*(2), 111–117.

Kotir, J. H. (2011). Climate change and variability in Sub-Saharan Africa: A review of current and future trends and impacts on agriculture and food security. *Environment, Development and Sustainability*, *13*(3), 587–605.

La Via Campesina and GRAIN. (2014). *Food sovereignty: Five steps to cool the planet and feed its people*. Retrieved June 11, 2021, from https://grain.org/article/entries/5102-food-sovereignty-five-steps-to-cool-the-planet-and-feed-its-people.

Simmons, D. (2019). *A brief guide to the impacts of climate change on food production*. Retrieved June 11, 2021, from https://yaleclimateconnections.org/2019/09/a-brief-guide-to-the-impacts-of-climate-change-on-food-production.

Soubry, B., Sherren, K., & Thornton, T. F. (2020). Farming along desire lines: Collective action and food systems adaptation to climate change. *People and Nature*, *2*(2), 420–436.

CHAPTER 15

The Future of Food

INTRODUCTION

Imagination is our greatest ally in tackling the challenges of the future of food production. While some ideas, such as floating, wind-powered greenhouses are still stuff of science fiction, the supertrees of Singapore, shown in Figure 15.1, are real. They are not trees but rather exhaust vents for the city. Combining art, technology and nature, the trees are the centerpiece of a 250-acre (100-hectare) botanical garden called Gardens by the Bay. They support climbing plants, generate solar power and collect rainwater. A partnership between Singapore's government and an architectural firm, this vision of a garden in one of the most densely populated cities in the world demonstrates that cities can be productive spaces and gardens on a massive scale and can provide ecological benefits for the city. This chapter concludes the book with a discussion of future trends and summarizes the connections between capitalism, environment and inequality and how current configurations of power in the food system hinder progress toward equity.

The central themes of this book focus on human environmental health, inequality and sustainability. Looking through the lens of place, region, distance, interconnection and interdependence permits geographers to see that what happens in one place affects what happens in another. In the case of the food system, development and health in one place can mean underdevelopment and inequality in another, or lack of sustainability in one place leads to lack of sustainability in other places. These patterns are driven by uneven access to and use of power by food system actors. In what follows, some lessons are taken from some current food trends and some ancient ones and trends in the political-economic governance of the food system and where we are headed or could be headed are discussed. The focus is on contemporary innovations in food production, agroecology, vertical farms and the decentralization of food and energy production and what that means for food consumption patterns as well as new organization of power within the food system. Also addressed is how the current configuration of power in the food system inhibits goals of equity and sustainability and how to change it.

DOI: 10.4324/9781003159438-19

FIGURE 15.1 Singapore supertrees and botanical gardens (https://commons.wikimedia.org/wiki/File:Supertree_Grove,_Gardens_by_the_Bay,_Singapore_-_20120712-02.jpg)

REVISITING THEMES

The central themes of this book emphasized sustainability as the intersection of social equity and justice, environmental health and economic viability. Two aspects of sustainability have been of interest to me in this book: social equity and justice and human and environmental health. The focus has been on these two themes of sustainability primarily because they receive less attention in general and also because social movements, like the environmental movement, are gaining ground in the fight for justice, and the food system is a place ripe for change. The focus has been less on the economic aspects of sustainability and more on the social and environmental dimensions primarily because of the overwhelming dominance of capitalism in our current system and the existence of very few sustainable alternatives. In this chapter, some time was spent thinking about economic alternatives, but first some connections must be drawn between these themes.

Few disagree that the contemporary, industrial food system is flawed and problematic in various ways. It produces a wide variety of foods for a small number of people (usually an elite class of White former colonizers), and the supply chains that produce those foods are rooted in colonialism and replicate past injustices in terms of labor (usually originating in a former colony). Chemicals and practices that we know are dangerous to humans and the environment continue to be used in spite of public outcry. The industrial food system produces and distributes too much food for one group of people in one place and not enough for people in another, simultaneously producing waste and hunger. Industrial agriculture also puts in more energy via petroleum than it puts out in food calories, make it an unsustainable system. Attempts to fix these problems have only reproduced, in new forms, the inequities that already existed. The Green Revolution reorganized the food system into an export model, and the sustainability turn produced a two-class food system.

The actors with the most power in the food system are *not* the most numerous: the consumers and producers. The real power in the food system lies with the actors with the fewest numbers. Their small numbers allow them to form bottlenecks in the supply chain to control what food goes where and in what kind of condition. The state regulatory agencies that govern food safety use policy to determine how food must be delivered from producers to consumers, who in the absence of a right to food, have very little power. We can see this in the geographic disparities in raw milk sales, for example. Those places that require pasteurization and criminalize the sale of raw milk have a small number of pasteurizing facilities, through which all milk must flow. This often prevents smaller-scale and agroecologically produced milk from being provided to consumers. This power struggle is also evident in access to seeds. Those companies that retain patents on seeds through state regulatory agencies control what farmers may grow and not grow and using what kind of practices. These practices are often harmful to the environment and to human health but are encouraged by state governments that have an interest in overproducing certain commodities.

The reason these problems persist in the food system is because they are rooted in capitalism. In short, there is money to be made from the suffering of laborers and from the use of dangerous chemicals. If there was not money to be made, if those with the power to do so structured the food system around a different set of priorities and needs, we could solve many of these problems. While it may be counterintuitive, the solutions to food system problems lie less with food production but more with the power dynamics that shape its conditions. These include the over-reliance on producing food as a commodity and the power of governments to control the food system through food safety, patents and trade agreements, to name a few. States and markets work together to overproduce certain commodities, which are then used as tools of foreign and domestic policy, rather than as ways to nourish people and sustain livelihoods.

Kate Raworth, economist and author of the revolutionary *Doughnut Economics* (2017), argues that the answer lies in changing the goal of the economy from a myopic focus on gross domestic product or capitalist growth, to "meeting the human rights of every person within the means of our life-giving planet" (22). The way to do this is to update our economic theory from flawed models of perpetual growth to one of an **embedded economy**, one in which the three sectors of social life, state, markets and civil society (which mirror the actors of food system governance) work together in balance toward the goal of meeting human needs for well-being. This would mean shifting power in the food system away from global markets where food is bought and sold as a commodity and returning it to food as a shared resource in a community.

An embedded economy would be one that prioritizes health, equity and well-being. It would also be one that values the household as a central part of the productive economy, along with the state, markets and the commons, which are also the primary actors shaping food system governance. These are embedded within society, which acts as a guardrail or a speed bump to the growth of the economy, the power of the state, the abuse of the commons and the exploitation of women's work within the household. Society is embedded within the life-giving bounds of the Earth, through which the energy of the Sun, but nothing else, passes through. Viewing the Earth as a closed system with finite resources but an unlimited amount of presently available solar energy is game changing and also deeply relevant to the way in which agriculture is practiced and food is produced.

Within such a system, food as a human right, not profit, is the goal. The household is a primary place of production that could be remunerated properly for its work (cooking, cleaning, caregiving, etc.) through a **universal basic income (UBI)**, which is a small monthly stiped designed to cover basic costs of living (Bregman, 2017). The state would use its power and resources to innovate and check the power of markets, which would provide goods efficiently and affordably, sourced sustainably from the commons and not privately owned land. Society would use its democratic power to influence policy, ensure that people are not hungry and set rules and make enjoyable, delicious culture around food. A system such as this, embedded in a solar economy, would rely on renewable energy and the paid labor of humans who would provide for each other in mutualistic *and* market-based ways. Discussed in what follows are some actually existing possibilities that are currently realizing such goals, at least in part.

FUTURE FARMS

Urban agriculture, a new concept as of the late 20th century, is a growing trend globally and will certainly be the future of food production. With more than half the world's population living in cities and with disasters threatening supply chains and the availability of labor, urban agriculture will not be a luxury but a necessity in the future. Countless examples from Paris to Havana of urban gardening using common land populate this book, including the garden featured in Figure 15.2 in Wellington, Australia. Urban gardens, particularly when situated on unused or unclaimed land, offer producers and consumers new forms of power in the food system. Combined with food forests, **vertical farming** in greenhouses that house solar power, rainwater catchment and aquaponic systems and composting of food scraps, urban gardens and farms hold the promise of providing high-quality, nutritional food for free or for very low prices. They also close energy loops by utilizing human labor, solar power and rainwater, and reduce waste. Many of these projects originate in food-insecure communities because of the lack of food retail options. It is a model of sustainable design for a low fossil fuel future and has the potential to transform the way other supply chains distribute food as well. Rather than relying on long, vulnerable supply chains, urban food production can, with the proper state support and consumer demand, decentralize to a more local and regional distribution system that complements but does not replace a more energy-intensive global food supply system.

The future of food will also expand what we eat to include things like aquaponic tilapia and sprouts, but also such food as algae and foods derived from fungus and yeasts and imitation meats. These foods all can be grown cheaply and/or sustainably and are extremely versatile. Algae is a versatile, affordable and nutritious food that can be harvested sustainably. It takes many forms, and we already consume it in a variety of ways

FIGURE 15.2 Urban farm in Wellington, Australia (https://commons.wikimedia.org/wiki/File:Kaicycle_Urban_Farm_Wellington_02.jpg)

through food additives. Lab-grown or imitation meat is certainly a way forward for some meat consumers if/when the production process can be simplified and made affordable. These foods may not replace what we currently eat, but they may begin to complement or replace globally sourced foods from the tropics and other processed food when the expense of fossil fuels makes their distribution cost-prohibitive. Some food crops may also be repurposed for industrial uses and replace traditional petroleum products. Plant-based plastics that can be composted and turned back into soil to grow crops to make plastics are closed, sustainable loops, and they are also on the rise. Hemp is known for its versatility as a manufacturing base and is a leading candidate crop for such an enterprise.

The production of actual food on a large scale for local communities to eat or to process into useful forms will require the **regionalization of food systems**. While the local and sustainable food movements showed how animal proteins and produce can be produced on a small scale for local markets, these efforts have proved inadequate for feeding larger populations and the most vulnerable within those local communities. These systems have also not been able to produce those foods that form the basis of most human diets: grains and legumes. The scale of production that is necessary to produce these food items affordably and reliably is inconsistent with the small-scale nature of local food supply chains. However, regional cooperatives could pool resources and land in a way that could make it possible to compete with global markets and produce the food with agroecological methods. Food could be processed and sold through regional food hubs that make marketing and manufacturing cost-effective and efficient for medium-scale producers.

The process of **recentralization**, or returning the farm or the farmers market to the center of the community, is a spatial practice of reclaiming land for the right to food. None of this can or should be accomplished without an accompanying practice of **decolonization**. Decolonization is not a metaphor for reparations or recognition (Tuck & Yang, 2012) but an active process of returning sovereignty over stolen land to Indigenous people and leaving the future of settlers in their hands. According to Daigle and Ramírez (2019), this also includes "an affirmative refusal of white supremacy, anti-blackness, the settler colonial state, and a racialized political economy of containment, displacement and violence" (80). Without freedom from premature death, incarceration or expulsion from the settler state, the right to food, what people eat and how it is produced is a secondary concern. In my other work, I proposed a new Land Act that provides for the redistribution of land to those who wish to produce food, not commodities, with Indigenous people as the first recipients. This is not a complete answer, or maybe not even the right one, but it is a place to start thinking about how we might collectively, collaboratively restore balance to a broken system.

For example, the green roof shown in Figure 15.3 is a collaborative project between the University of Georgia (UGA) and the city of Athens, Georgia. Originally installed on the building in the 1960s when it was constructed to provide an ambient temperature environment for weather data collection and climate research, the green space now produces food. The food, primarily vegetables but also herbs and fruits, is grown in cooperation with the campus organic garden (UGArden) by an intern from the Office of Sustainability and students in an advanced undergraduate class. The UGA chapter of Campus Kitchens converts the vegetables and fruits into meals for a food-insecure population in the city: Grandparents Raising Grandchildren. Combining the resources of the non-profit sector, the social welfare arms of the city government and the labor of volunteers, food can be

FIGURE 15.3 Green roof garden on the Geography and Geology Building at the University of Georgia

provided to people without commodifying it. It is these new forms of partnerships that can overturn some of the negative power imbalances in the global food system that produce waste on one hand and hunger on the other.

FUTURE FORMS

To make all this good stuff happen takes reorganization of our economic systems and leadership from organizations, states and the private sector. It will require us to organize our economy, including our food supply chains, within our environmental and social justice goals. The **doughnut model of sustainable development** illustrates how there are environmental limits to what we can do with the Earth's resources without doing long-term damage: a key idea of sustainability. The model is two circles, one embedded inside the other, forming a space inside, (the doughnut hole). The areas outside the doughnut are zones of overshoot, where environmental conditions stray from the optimal model, including the loss of biodiversity, global temperature increases and pollution. On the inner boundary, in the doughnut hole lies the social justice foundation where shortfalls appear in the form of food insecurity, homelessness, disease and violence. In some cases, the boundaries are not quantified, but in others, such as with climate change and food security, we know very well where the tipping points lie. The Netherlands have implemented this model in development plans to rebuild the economy of Amsterdam in the post-pandemic era.

One important piece to achieving the well-being of communities is the idea of UBI. UBI is direct payment of a modest amount of money each month that can provide the basic necessities, such as food, housing and childcare for working parents. There are dozens of experiments ongoing around the world, including one launched in 2021 to eradicate child poverty in the US. Research has shown that direct payments are cheaper than caring for the homeless and the long-term consequences of poverty. The fears that UBI will encourage people not to work are largely unfounded, and many experiments show that the opposite was true. In Britain, money given to single parents *allowed* them to work and go back to school because they could pay for childcare. UBI also could eliminate food insecurity by giving people enough money to buy adequate food for themselves and their families. In his book *Utopia for Realists*, Rutger Bregman (2017) advocates for a 15-hour workweek in addition to UBI, citing the health benefits of reduced working hours, the benefits of full employment and the increased contributions to civil society from people who have more time for volunteerism.

The Combahee River Collective, a group of Black activists who formed in the 1970s, argue that "if Black women were free, it would mean that everyone else would have to be free since our freedom would necessitate the destruction of all the systems of oppression" (Taylor, 2017, p. 10). Women of color face multiple and interlocking forms of oppression centered on white supremacy, racial capitalism and heteropatriarchal social relationships and often have the least amount of power in the food system. Raworth (2017) identifies the lack of remuneration for the unpaid work of the household, often taking the form of food production, procurement and processing, and almost universally done by women, as a key flaw of economic models. When women are compensated for the work in the household, all of society benefits: increased capacity for childcare, better and healthier food and more time for the work of civil society. A UBI has been shown to be one of the single most significant contributions to the reduction of poverty and food insecurity. Combined with access to universal healthcare and reproductive freedom, the improvement to women's lives has the capacity to restore balance and health to populations. An activist demonstrating at the US presidential inauguration in 2021 shown in Figure 15.4 is hopeful that the experimental reforms to work (i.e., income support, unemployment, child payments) and healthcare (i.e., free testing, vaccinations and medical care) in the context of the COVID-19 pandemic may be realized more widely in society. These reforms are central to the reconfiguration of power in the food system because they alleviate the poverty that causes hunger and provide farmers with the necessary healthcare to pursue their livelihoods.

Raworth (2017) identifies the separation of workers from owners as a source of unsustainability in the capitalist system of production. Providing people, however skilled, with a UBI will free them from a wage-labor relationship that exposes them to exploitation and abuse. It also frees people to produce their own food and to join cooperatives to produce food that then does not necessarily need to be sold to be distributed, thus freeing food from its commodity form. A model for larger-scale more capitalist enterprises, however, is

FIGURE 15.4 Demonstration on the US Presidential Inauguration Day in 2021 (https://commons.wikimedia.org/wiki/File:Inauguration_Day_2021-_Columbus_(1-20-21)_02eIMG_7596_(50871896272).jpg)

the Mondragon Corporation, which is a worker-owned cooperative in the Basque country in Spain. It was founded in 1956 as a way to foster participation and solidarity in workplaces. The cooperative is organized around a few key principles of social justice, anti-capitalism and common goods. At the center of its value system is education and employee ownership. Derived from this are wage solidarity, participation in management, subordination of capital to labor, internal accountability among workers and non-discrimination.

Wage solidarity means that there is a small differential between the lowest paid and the highest paid worker. In the case of Mondragon, this ratio is 1:6. The **subordination of capital to labor** means that worker well-being is more important than profit. While it has not been without its problems, particularly during the 2008 financial crisis, the cooperatives were at the heart of some of Spain's only successful economic regions. During the hardest times of the economic crisis, while the rest of Spain endured 26% unemployment, the cooperatives maintained steady employment and production. During the COVID-19 pandemic, the agri-food co-ops were able to help other agricultural industries in Spain retool their supply chains to meet new demands for food in retail outlets instead of restaurants. They also hired unemployed citizens who could work without losing their unemployment benefits, in an unofficial form of basic income.

None of this is possible without strong democratic institutions that respond to the demands and needs of the people. Direct democracy, which means that all citizens vote on every decision a government makes, rather than electing representatives to do that work for the population has, for various reasons, not been a popular choice for governing in recent years. Another reason is that it returns power to people who can have a greater say in what shapes their lives, rather than private capital and the nation-states that support it. Another reason is purely logistical. When the population numbers in the millions, it is difficult to organize people to vote often. The internet party, however, offers a solution. Vote on your smartphone. Vote multiple times per day. We do this already in so many ways. In a literate population with access to technology, the internet offers a real path toward direct democracy. Several countries in Europe already have well-established internet parties and are gaining in popularity.

BLUE ZONES

The doughnut model of economics stresses human well-being and flourishing, so looking for places where that happens can tell us a lot about how to do it. **Blue Zones** are longevity hotspots in the world where a disproportionately large percentage of people live well into their 100s without debilitating diseases (Buettner, 2012). Researchers identified five such Blue Zones in the world: Okinawa, Japan; Nicoya, Costa Rica; Loma Linda, California; Sardinia, Italy; and Icaria, Greece. Researchers set out to discover what might be driving such long, healthy lives. They narrowed it down to four main factors, one of which, obviously, is a healthy diet. The other three factors are just as important as what people ate and cannot be considered in isolation. They all spent time outdoors walking, gardening, fishing or doing other hobbies, and one can presume not working in an office doing sedentary work. They all belonged to a tight family group or something that resembled a big extended family, such as a religious community. They also had a philosophy of life that helped them deal with stress, whether it was faith in a higher power or a sense of purpose in life. Figure 15.5 shows the four factors and where they overlap to generate well-being and health over the long term.

The dietary dimensions of a Blue Zone are interesting because they mimic what researchers have found to be sustainable in terms of land use and ecosystem health. The majority of food in their diets was plant-based, including greens, lentils, starchy roots, fruits and vegetables, but most also consumed a small amount of animal-based products. Animals are beneficial in natural ecosystems, and managing the population through selective

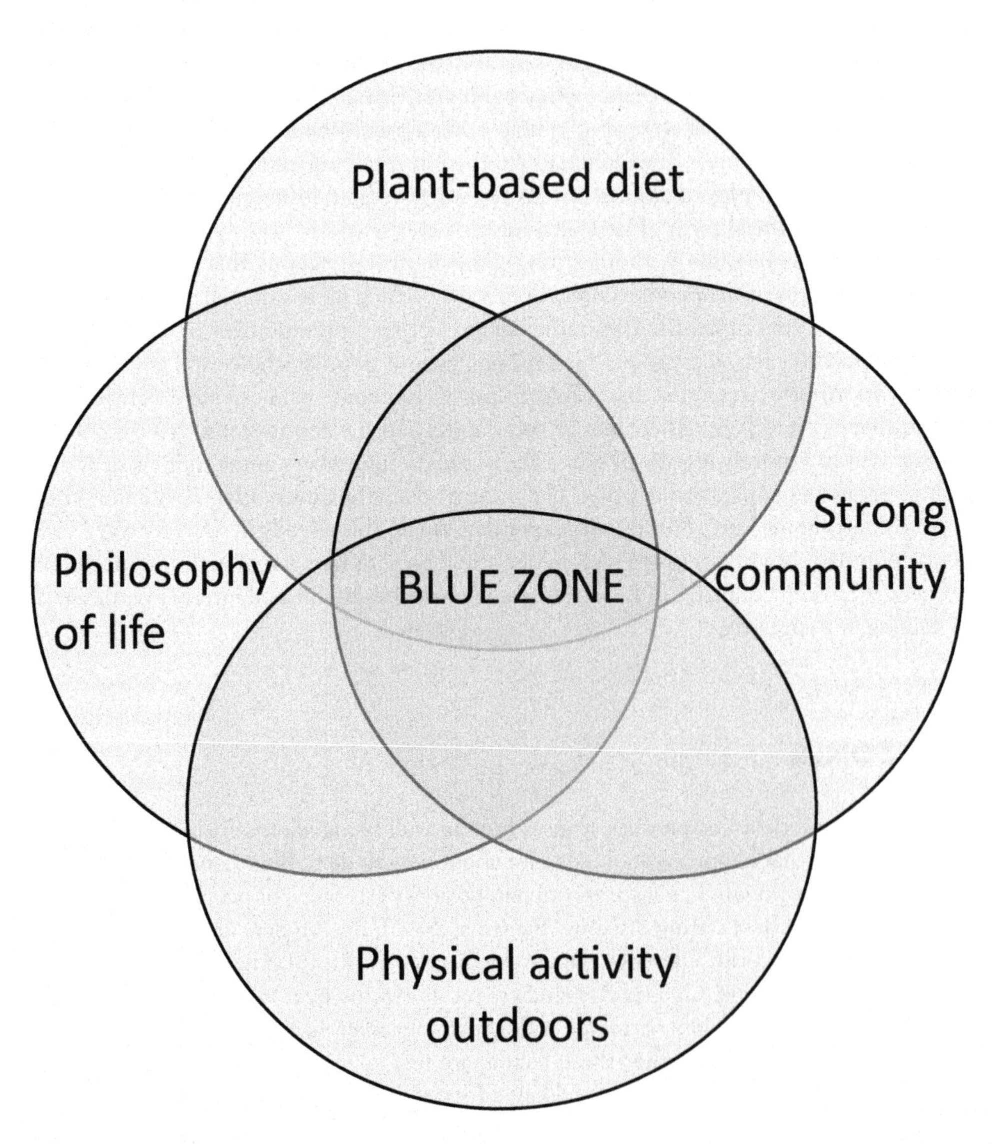

FIGURE 15.5 Blue Zones

harvest is a sustainable practice. The researchers put the four factors into a pyramid with the feeling of belonging, family and community at the bottom, with a diet based on plants, and eating only until you are 80% full. The Blue Zone researchers stress that eating with others who are part of your strong social network and/or growing food with others is a vital piece of the puzzle. Having a purpose and reducing stress follow, and the pyramid is topped with natural, daily movement, ideally outside. Anecdotes from the research show people enjoying growing food, making wine and eating together well into their 10th decade. It does not seem possible to enjoy all this well-being while working 40 hours/week, so maybe Bregman's 15-hour workweek is essential to health and longevity as well. A regional food system could provide the means for producing these foods for a healthier diet as well as provide meaningful work for members of the community.

SUMMARY

Some of the most powerful tools to change the food system have nothing to do with food or agriculture. With UBI, a 15-hour workweek, **worker-owned enterprises**, urban agriculture, **direct democracy** and a sustainable model of development, new configurations of power in the food system make possible a healthy, fair and delicious future. While the agenda of decolonization, decentralization and alternative models of development are the work of states and civil society to do together, there are things individuals can do, starting right now. Seek out sources of food grown without chemicals and with fairly compensated, free human labor. Be close enough socially and physically to know. Volunteer at a farm to learn more about it and to help build a reciprocal economy. Start growing your own food. A basil plant in a sunny window is easy to do and will save plastic and fossil fuels. Instead of reaching for the airlifted pineapple or mango in the supermarket, reach for an apple, peach or pear, or something more suited for your latitude. Or seek out a local source of fruit and learn how to process it when it is out of season. Let pineapple be food for people where pineapple grows. Let farms return to the center of our communities. Take power from global governing actors and return it to communities. Let flourishing be a reality for all.

SUPPLEMENTAL MATERIALS

Key Terms and Concepts: Blue Zone, decolonization, direct democracy, doughnut model of sustainable development, embedded economy, recentralization, regional food system, subordination of capital to labor, universal basic income, urban agriculture, vertical farming, wage solidarity, worker-owned enterprises

Explore More: Kate Raworth is an economist who identified a model for producing social equity and justice while preserving environmental resources and health. Explore more here: www.kateraworth.com/doughnut/

Recommended Viewing: *A Farm for the Future* (2009)

Discussion Prompt: What do you think is the future of food? Of all the topics we covered this semester, what stands out to you as the biggest issue in the food system

and how should we tackle it? Aim for 100–250 words. Be sure to correctly use at least *one concept* from this chapter in your discussion post and to write it in **bold** font.
Recipe: Blue Zone foods include lots of vegetables, whole grains and natural movement outdoors. A recent article in my favorite cooking magazine *Eating Well* put a portable spin on sushi with whole grain rice and smoked salmon wrapped in seaweed that can be packed to go. Add some edamame and raw vegetables and you've got a healthy picnic for a hike in the woods to start living the Blue Zone life. www.eatingwell.com/recipe/7902714/smoked-salmon-brown-rice-onigiri/

REFERENCES

Bregman, R. (2017). *Utopia for realists: And how we can get there*. Bloomsbury Publishing.

Buettner, D. (2012). *The blue zones: 9 lessons for living longer from the people who've lived the longest*. National Geographic Books.

Daigle, M., & Ramírez, M. M. (2019). Decolonial geographies. *Keywords in Radical Geography: Antipode, 50*, 78–84.

Raworth, K. (2017). *Doughnut economics: Seven ways to think like a 21st-century economist*. Chelsea Green Publishing.

Taylor, K. Y. (Ed.). (2017). *How we get free: Black feminism and the Combahee River Collective*. Haymarket Books.

Tuck, E., & Yang, K. W. (2012). Decolonization is not a metaphor. *Decolonization: Indigeneity, Education & Society, 1*(1).

Index

Page numbers in *italic* indicate a figure on the corresponding page.